MEGAWATT INFRARED LASER CHEMISTRY

MEGAWATT INFRARED LASER CHEMISTRY

ERNEST GRUNWALD
Brandeis University

DAVID F. DEVER
Macon Junior College

PHILIP M. KEEHN
Brandeis University

A WILEY-INTERSCIENCE PUBLICATION

JOHN WILEY & SONS

New York • Chichester • Brisbane • Toronto

Library of Congress Cataloging in Publication Data:

Grunwald, Ernest
  Megawatt infrared laser chemistry.

  "A Wiley-Interscience publication."
  Bibliography: p.
  Includes index.
  1. Lasers in chemistry. 2. Infra-red sources.
I. Dever, David F., joint author. II. Keehn, Philip M., joint author. III. Title
QD715.G78 541′.353′2 78-6721
ISBN 0-471-03074-0

Printed in the United States of America

10 9 8 7 6 5 4 3 2 1

A good inductive model should have several attributes: It should be intentionally incomplete. The possible disproof of an inductive theory [is] not a flaw in the method, but... an essential and welcome part of the process of understanding. The model should be able to be refined.... It is poor inductive practice to foreclose the possibility of later insertion of new knowledge. Calibration [of the model] to experiment is essential.... It is the process of calibration and subsequent use that makes the heaviest demands on one's chemical judgment.

Don L. Bunker (1931–1977)

Preface

During the past two or three years we have found research in infrared laser chemistry engrossing, not only because of the practical aspects and prospects of this budding science, but also because the required theory reaches to the very foundations of chemical dynamics. This sentiment is shared by others, for research is in a stage of rapid growth and major chemical industries are thinking seriously about large scale applications.

The volume is intended as a greeting to new workers in the field, and as a primer to acquaint chemists in general with the salient features of the phenomenon. For this reason the contents have been limited largely to established facts, theories, and basic issues. The contents have also been limited largely to chemistry done with pulsed infrated lasers at megawatt per square centimeter power levels, which (it seems to us) is the most important branch of infrared laser chemistry about which a book that addresses itself to chemists can be written at this time. At gigawatt per square centimeter pulsed power levels where laser-isotope separation was first discovered, the chemistry still seems very complicated, being best described as molecular fission rather than as orderly decomposition. At the more modest power levels of continuous infrared lasers, controversy surrounding even characterization of the phenomenon is still too widespread.

Throughout the book we have carefully avoided riding theoretical hobby-horses, even those from our own stables. However, we have felt no similar compunction about drawing on published data from our own laboratories for illustrative examples.

It is a pleasure to acknowledge helpful discussions with many individuals, particularly with our students and collaborators C. Cheng, D. Garcia, G. A. Hill, C. M. Lonzetta, and K. J. Olszyna. E. G. expresses special thanks to H. J. Wetzstein for having aroused his interest in this subject and for steady counsel. D. F. D. acknowledges the help of L. Fountain,

R. Land, S. Rusk, and D. White with reference material. We thank Ethel Crain for typing the final manuscrpt and James C. Magno for the artwork.

Ernest Grunwald
David F. Dever
Philip M. Keehn

Waltham, Massachusetts
Macon, Georgia
May 1978

Contents

Glossary of Symbols

The units commonly used in this book are given in parentheses.

c	speed of light
CPF	conversion (of reactant) per flash (fractional, or %)
CW	continuous wave
D	dose (radiant energy per unit area); dose transmitted by sample; also called *energy fluence* (J/cm^2)
D_0	dose incident on sample (J/cm^2)
e_A	extinction coefficient by laser spectroscopy $(\text{cm torr})^{-1}$, Eq. 2-6
E_{abs}	energy absorbed per mole of substance (kcal)
E_{act}	Arrhenius activation energy (kcal)
E_0	amplitude of electric vector of coherent radiation
ε_A	extinction coefficient by conventional spectroscopy $(\text{cm torr})^{-1}$, Eq. 2-5
$\tilde{\epsilon}$	rms-average energy exchanged with heat bath, per collision
f	spoil factor
f_k	mole fraction of molecular oscillators in kth level
GW	gigawatt, 10^9 watt
I	beam intensity (radiant power per unit area); intensity transmitted by sample (W/cm^2)
I_0	beam intensity incident on sample (W/cm^2)
J	rotational quantum number
K	collisional efficiency parameter, Eq. 3-7
l	path length
MW	megawatt, 10^6 watt
$\lvert\mu\rvert$	electric dipole transition moment
N_0	Avogadro's number
ν	frequency; wavenumber (cm^{-1})
ω_R	Rabi frequency
p	probability
P	pressure (torr)

P_A	pressure of absorbing gas (torr)
P_{eff}	effective pressure of gas mixture (torr)
P_X	pressure of nonabsorbing gas in mixture (torr)
s	number of oscillators
TEA	transversely excited, atmospheric pressure
τ	effective pulse duration, Eq. 3-4
v	vibrational quantum number
$\bar{v}$	ensemble average of v for the given vibrational mode
Z	gas-kinetic collision number per second, under standard pressure conditions

MEGAWATT INFRARED LASER CHEMISTRY

Chapter 1

Introduction

Infrared photons are relatively energy poor. For instance, a mole of photons at 1000 cm^{-1} carries only 2.86 kcal. It was therefore a pleasant surprise when it was discovered a few years ago that exposure of gases to high-power infrared laser beams could induce chemical reaction [1]. Previous to this, photolyses were done with visible, ultraviolet or even shorter wavelength radiation.

Infrared photochemistry is a new kind of photochemistry because reactions take place in the electronic ground state. It offers the potential of high specificity because the infrared laser radiation is monochromatic and can be tuned to a frequency which matches an absorption frequency of the molecules to be excited. Because the molecules can "climb up the vibrational ladder," absorption is not limited to one photon per molecule. Indeed, with commercially available pulsed lasers, it is easy to introduce 10 kcal/mole or more with each pulse when the laser is tuned to a strong absorption band.

From a practical point of view, it is convenient to divide laser chemistry into three subdivisions, depending on the power level used. When excitation takes place with continuous-wave (CW) lasers, the power is necessarily limited to a few kilowatts per square centimeter or less; continuous absorption at higher power levels would atomize the molecules. When excitation takes place with pulsed lasers, the outputs per pulse of commercial tunable lasers are typically 1 J/cm^2, with pulse durations ranging from 0.2 to 1 μs. This corresponds to average power levels of 1–5 MW/cm^2. The beams produced by such lasers have radii up to several centimeters. This kind of radiation is therefore suitable for studies involving fairly large samples. Finally, when megawatt beams are concentrated by means of lenses, power levels of gigawatts (GW) per square centimeter are obtained near the focal point. Gigawatt power levels have been found useful for isotope separation work under "collision-free" conditions, mostly at pressures below 1 torr [2–4].

In this volume we consider almost exclusively megawatt laser chemistry, which we believe to be of special interest to chemists. First of all, pulsed megawatt irradiation is a convenient way to study high-temperature ($>1000°K$) reactions in what is essentially room-temperature apparatus. The effective temperature is readily changed over wide ranges by changing the infrared dose. This permits the detection and detailed study of reactions heretofore inaccessible because of their high activation energies.

Also, reactions observed in laser chemistry appear to be cleaner than their counterparts studied in classical thermal apparatus for several reasons: The reactant is exposed to high temperature conditions for only a short length of time. The initial "heating rate" is in the range of 10^6–10^{11} °K/s; the subsequent "cooling rate" is in the range of 10^3–10^6°K/s. If the laser beam is smaller than the reaction cell cross-section, wall effects are practically absent. If the laser beam irradiates not only the gas but also the cell wall, reactions occurring on the cell wall are accentuated.

Because high-temperature reactions frequently have bond-breaking steps as primary processes, very high concentrations of free radicals (>10 torr) can be generated in a matter of microseconds and studied by kinetic spectroscopy. Contrary to conventional free-radical chemistry where bimolecular reactions *among the free radicals* are stoichiometrically unimportant, the high concentrations of free radicals produced by laser flash lead to dominance of radical-radical reactions.

The primary steps of laser-induced reactions in most cases are decomposition reactions. However, examples are accumulating in which the original molecules are being isomerized, or in which larger molecules are being synthesized by bimolecular combination or displacement.

Chapter 2

Some Phenomenology

One of the perquisites of participating in the development of a new field is being able to observe one of its symptoms: theories come and go like the blooms on a rosebush. Infrared laser chemistry exhibits this symptom actively; therefore, until theories become more settled, this chemistry's claim to usefulness must be based on facts. In this chapter we consider some of the salient phenomena.

THE CO_2 LASER

At present the CO_2 laser is best suited in terms of energy, power, availability, and ease of use. It consumes only environmentally acceptable, commercially available supplies and is a clean and low-maintenance instrument operating in the 9.2–10.9 μ range. Bonds that absorb in this region include C—C, C—O, C—F, Si—F, S—F, P—O, W—O, and S—H. Other available lasers are the HF/DF laser (2.8–4.0 μ) and the CO laser (5.1–5.6 μ), but unlike the CO_2 laser, the gases they use (in necessarily large quantities) are either corrosive or lethal. These and other problems, involving monochromaticity or power, make their adoption for day-to-day use in the laboratory problematical at this time [5,6].

Nirvana in this field, of course, is a laboratory with a laser beam continuously tunable to any wavelength desired with high power and high efficiency. Any infrared-active vibrational mode could be stimulated, including those modes that absorb strongly. Actually, this may be not far from reality because there is now under development a laser tunable from 10 μ to 1000 Å; this covers the spectrum from the infrared stretch region to the vacuum ultraviolet [7]. To gain this high degree of tunability, a synchrotron is coupled to a low power laser; frequency selection is done by changing the energy of the electrons in the synchrotron. Present power levels are up to 10 kW and the investigators are discussing 0.1 MW as a reasonable goal for output for this instrument.

Emission Mechanism

The instrument used by the authors in their studies is a TEA (transversely excited, atmospheric pressure) tunable pulsed CO_2 laser with average power output between 0.1 and 2 MW/cm^2. Up to 10 J can be delivered in pulse times between 250 and 800 ns; the time between pulses is adjustable between 0.5 and 5 s. Beam areas are on the order of 8 cm^2 and doses* are reproducible to 2% in successive pulses. Its power supply and associated electronics are powered by 110 V AC, and it consumes gases at a rate of 5 ft^3/h. The physical arrangement for the laser consists of gas flowing through an 8 ft glass tube fitted with internal electrodes across which electrical energy stored in capacitors can be discharged. Optics appropriate to laser behavior are mounted at the ends of the tube. Photographs are shown in Figs. 2-1 and 2-2.

The dominant processes that occur in the CO_2 TEA laser are (1) molecules are excited to the upper vibrational laser level by collision with electrons from the discharge, (2) the molecules emit laser radiation, falling to the lower laser level, and (3) they then return to the ground state where they can start the cycle again [5]. In order to effect this cycle with high efficiency, N_2 and He are added to the CO_2; the N_2 assists in the first step, He in the last. The process can be expressed by a series of equations:

1. Excitation by electric discharge

$$N_2 + (e^-)^{**} \rightarrow N_2^{**} + e^- \tag{2-1}$$

*Laser physicists sometimes use the term *energy fluence* for dose and *energy flux* for intensity.

Fig. 2-1 View of the authors' infrared laser laboratory. The pulsed CO_2 laser is set up inside the Faraday cage (on the left) in order to contain electromagnetic noise. Wide-angle lens photograph by John Galano (1977). Courtesy Brandeis University.

Fig. 2-2 Inside view of the CO_2 laser. Excitation and stimulated emission take place in the long glass tube that is mounted alongside the high-voltage condensers. There is some optical distortion owing to wide-angle lens photography. Photograph by John Galano (1977). Courtesy Brandeis University.

2. Collisional transfer to upper laser state

$$N_2^{**} + CO_2 \rightarrow CO_2^{**} + N_2 \tag{2-2}$$

3. Laser emission, falling to lower laser state

$$h\nu + CO_2^{**} \rightarrow CO_2^{*} + 2h\nu \tag{2-3}$$

4. Return to ground state

$$CO_2^{*} + He \rightarrow CO_2 + He^{*} \tag{2-4}$$

An energy schematic with each of the above steps labeled is shown in Fig. 2-3. It will be noted that there are two lower laser levels, one at 1388 cm^{-1} and one at 1286 cm^{-1}. These help to give rise to the wide selection of laser emission lines available to the researcher. The latter lines are indicated in the bottom part of the figure.

One of the tube-end optics mentioned above is a grating with which the laser can be tuned to any one of the emission maxima. This high degree of choice arises from the fact that the upper and lower laser levels

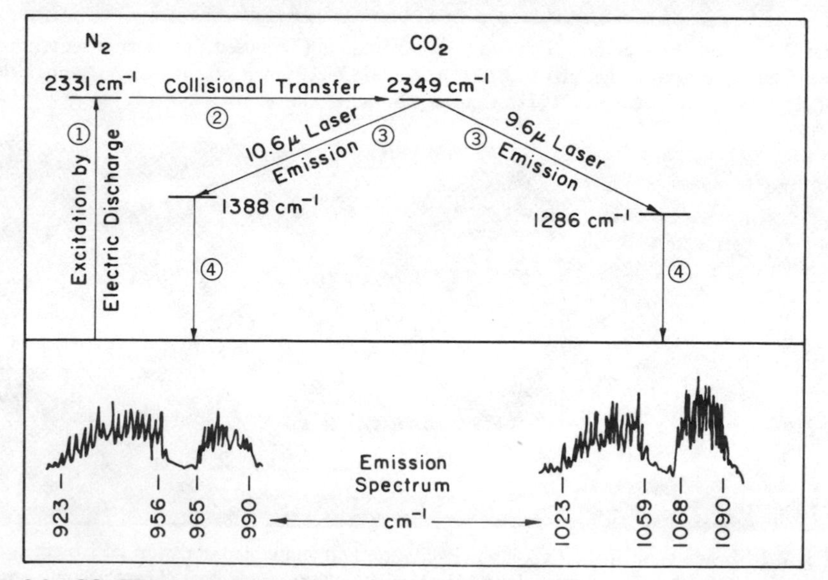

Fig. 2-3 CO_2 TEA laser excitation scheme. Circled numbers refer to step-sequence discussed in text. Emission spectrum shown below laser transitions indicates tuning range of laser.

each have rotational levels (not shown in the figure), about 40 of which have substantial population densities [5]. It is between pairs of these levels (selection rules permitting) that transitions take place. Tuning the laser to resonate on any one of these lines gives the operator control of the laser emission frequency.

ABSORPTION AND REACTION

For effective absorption of radiant energy, it is desirable that the laser be tuned so that its output frequency falls into the range of a strong absorption band of the substance to be excited. By "strong" we mean that the extinction coefficient, ε_A, of the absorbing species [defined in (2-5) and measured by ordinary nonlaser spectrophotometry] must be at least of the order of 0.001 cm^{-1}/torr at the band maximum.

$$\log_{10}\left(\frac{I_0}{I}\right)=\varepsilon_A l P_A \quad (2\text{-}5)$$

Multiphoton absorption from megawatt laser sources is a complex phenomenon and does not in general follow the Beer-Lambert law [8,9]. Nevertheless, it is useful to introduce a quasi-extinction coefficient, e_A, by way of (2-6), in which D_0 and D denote the incident and transmitted dose, respectively; that is, D_0-D is the amount of radiant energy absorbed by the gas per square centimeter.

$$\log_{10}\left(\frac{D_0}{D}\right)=e_A l P_A \quad (2\text{-}6)$$

Because the real extinction coefficient, ε_A, is only an imperfect model for e_A, the frequency at which e_A is a maximum may be different from that of the spectroscopic band maximum; the frequency at which e_A is at a maximum may also vary with dose. An approximate rule may be stated as follows [2]: Let ν_0 be the fundamental absorption frequency of the mode to be excited, and ν_1 be that of the first overtone. Owing to the anharmonicity of molecular oscillators, ν_1 is probably somewhat smaller than $2\nu_0$, a typical value for $(2\nu_0-\nu_1)$ being approximately 10 cm^{-1}. The rule then states that e_A will be at a maximum when the laser frequency is $\nu_1-\nu_0$.

Because $(\nu_1-\nu_0)<\nu_0$, for best absorption the laser frequency should fall into the P branch of the infrared absorption band.

To illustrate some of these points, Table 2-1 shows selected data for the infrared laser chemistry of CCl_3F [9]. When 60 torr of CCl_3F is irradiated at 1079 cm^{-1} in the P branch of the C-F stretching band, laser-induced decomposition takes place by a carbene mechanism:

$$CCl_3F \xrightarrow[E_{act}\sim 81\ \text{kcal/mole}]{nh\nu} :CClF + Cl_2 \tag{2-7}$$

The isomeric dichlorodifluoroethylenes are the only products detected.

$$2\ :CClF \rightarrow 0.43_5\ \text{(Cl)(F)C=C(Cl)(F)} + 0.43_5\ \text{(Cl)(F)C=C(F)(Cl)} + 0.13\ \text{(Cl)(Cl)C=C(F)(F)} \tag{2-8}$$

Table 2-1 illustrates the following points:

1. The extinction coefficient, e_A, varies slightly with dose. In this example, e_A decreases with increasing dose. However, for some substrates under certain conditions, e_A is a constant or increases with dose.
2. The magnitudes of e_A and of the spectrophotometric extinction coeffient ε_A are similar. In the present case, e_A is the smaller one; in other cases, the situation is reversed [10]. The relationship of e_A to ε_A varies with external conditions [10] and with the absorption band being irradiated [11].
3. Average amounts of absorbed energy (E_{abs}) are equivalent to several photons per molecule. In Table 2-1, values of E_{abs} range

Table 2-1 Selected Data for Infrared Laser Chemistry of CCl_3F at 1079 cm^{-1} [9]

$\bar{D}$ (J/cm^2)	e_A ($torr^{-1}/cm$)	E_{abs} (kcal/mole)	E_{abs} (photons/molecule)	CPF (%)
0.372	0.00332	13.3	4.3	2.8
0.270	0.00387	11.2	3.6	0.80
0.230	0.00403	9.9	3.2	0.40
0.135	0.00473	6.9	2.2	0.009
—	$\varepsilon_A=0.0135$	—	—	—

from 7 to 13 kcal/mole which, though important, are far less than the reaction activation energy of 81 kcal/mole. Nevertheless, considerable laser-induced reaction occurs per flash. This indicates that the molecular energy increments are nonuniform so that the observed values of E_{abs} per mole are in fact mean values for statistical distributions. The distribution functions are broad enough so that some, indeed many, of the molecules gain energy increments that are well above the mean and reach energy levels that lie above the activation barrier for reaction.

Percent conversion per flash (CPF) in successive flashes is approximately constant. Thus relatively small numbers of flashes are sufficient to give analytically significant amounts of product.

4. CPF is highly sensitive to E_{abs}. The data are reproduced well by the empirical equation (2-9), in which E_{abs} is expressed in kilocalories per mole.

$$\mathrm{CPF} = 1240 \exp\left(\frac{-81.1}{E_{abs}}\right)\% \tag{2-9}$$

Note that the energy parameter, 81.1 kcal/mole, is practically equal to the thermal activation energy for the primary decomposition step, (2-7).

Photosensitized Experiments

Because continuously tunable lasers are not yet available, there are many substances whose infrared laser chemistry cannot be excited directly because they do not have strong absorption bands in the CO_2 laser's tunable range. Such substances can be activated conveniently, though with lower efficiency, by the addition of an infrared sensitizer. The following example illustrates the technique.

The thermal rearrangement of allene to methyl acetylene is of mechanistic importance because cyclopropene has been suggested as a short-lived intermediate [12].

$$H_2C{=}C{=}CH_2 \rightarrow \left[\text{cyclopropene: } \mathrm{CH_2} \text{ bonded to } \mathrm{HC{=}CH}\right] \rightarrow CH_3C{\equiv}CH \tag{2-10}$$

Because the infrared spectrum of allene does not have a suitable, strong absorption band, the reaction cannot be induced by direct absorption from a CO_2 laser. However, when silicon tetrafluoride is added and the laser is tuned to 1025 cm^{-1} where SiF_4 absorbs strongly, rearrangement is observed when the mixture is irradiated [13]. SiF_4 is a suitable sensitizer not only because of its very strong infrared absorption but also because of its extraordinary chemical inertness [8]. Sulfur hexafluoride has also been used as an infrared sensitizer, and is satisfactory for low-power CW irradiation [14]. However, at megawatt power levels SF_6 is not generally inert, but reacts vigorously with hydrogen and with substances having C-H bonds [8].

PRACTICAL SEPARATION OF VARIABLES

The numerous CO_2 TEA laser parameters that can be adjusted include frequency, dose, average intensity, and pulse shape. Varying them during a series of experiments does indeed have an effect on CPF, the variable we use to follow the course of a laser-induced reaction. However, in our experience with MW laser pulses, the effect is felt only to the extent that such variables modify E_{abs}: simple knowledge of E_{abs} is sufficient for quantitative prediction of CPF in a system of constant composition and pressure.

Thus it is possible to separate the overall problem of CPF as a function of laser parameters into two separate problems: CPF as a function of E_{abs}, and E_{abs} as a function of laser parameters. This phenomenological separation permits a great simplification and has important theoretical implications.

For instance, the infrared flash photolysis of $CHClF_2$ has been studied in detail and appears to proceed by the following mechanism:

$$CHClF_2 \underset{}{\overset{nh\nu}{\underset{E_{act}=56\ \text{kcal/mole}}{\rightleftharpoons}}} :CF_2 + HCl \tag{2-11}$$

$$2:CF_2 \rightarrow C_2F_4 \tag{2-12}$$

Table 2-2 illustrates the practical independence of CPF of pulse duration τ under otherwise fixed conditions. E_{abs} is measured within 10% accuracy, while CPF is measured within 15%. Since a 10% change in E_{abs} would

Table 2-2 Effect of Infrared Pulse Duration on CPFin the Flash Photolysis of $CHClF_2$ [10]

P (torr)	E_{abs} (kcal/mole)	τ (ns)	CPF (%)	Z_τ[a]
$50/CHClF_2$	11.9	270	2.1	177
		800	2.6	524
$9.4/CHClF_2$	11.8	270	1.6	33
		800	1.7	99
$9.4/CHClF_2$	16.1	270	11.6	33
		800	8.8	99
$50/CHClF_2$	22.6	270	0.11	850
$+250/N_2$		800	0.11	2500

[a]Mean number of collisions experienced by a $CHClF_2$ molecule during the time τ.

cause a 40% change in CPF, the small, apparently random, variations in CPF with τ in Table 2-2 are statistically insignificant. On the other hand, the relation between E_{abs} and infrared dose does seem to depend on τ. For a fixed dose, more energy is absorbed from longer pulses, presumably because more time is available for the required microscopic processes.

THERMAL VERSUS PHOTOCHEMICAL MODELS

Since the pioneering work of Eigen, the technique of temperature jump (T-jump), followed by chemical relaxation to a new state of chemical equilibrium, has become a familiar method for the study of fast reactions. When an infrared laser chemist uses the term *thermal* reaction, he means the following analogous sequence of events:

1. The radiation absorbed from the infrared pulse is converted rapidly into random heat energy, producing a T-jump.
2. Normal high-temperature reactions take place while the gas is cooling back down to room temperature.

In this model, the amount of reaction during the time of the T-jump is considered negligible.

When an infrared laser chemist uses the term *photochemical* reaction, he is speaking of a reaction that takes place prior to the completion of the *T*-jump. The molecular energy distribution among the reactive molecules during this event is necessarily not at thermal equilibrium.

The processes of thermal reaction and photochemical reaction are not mutually exclusive. For instance, in the case of some infrared laser-absorbing substances, it is possible that some reaction takes place during, and some after, the *T*-jump. For mixtures of reacting substances, it is possible that the energy-absorbing substance reacts photochemically, while other substances react after the *T*-jump. We now consider experiments for distinguishing between these modes of reaction.

Dependence of CPF on E_{abs}

For thermal reaction, CPF in a given system depends only on the *magnitude* of the *T*-jump and is independent of the *T*-jump dynamics. Thus, for thermal reaction, CPF in a given system depends only on the amount of absorbed energy and is independent of such jump-dynamics variables as the duration of the infrared flash, the nature of the absorbing vibrational mode, and (for mixtures) the nature of the absorbing molecular species. However, the condition that CPF depend solely on E_{abs} is not a sufficient condition. It is conceivable, for instance, that the absorbed energy spreads rapidly among the vibrational degrees of freedom, reaching a stationary state in which vibration is not yet in thermal equilibrium with translation and rotation.

Cooling Rate

When CPF in a given system is found experimentally to depend only on E_{abs}, a test for thermal reaction may be made by deliberately changing the rate of cooling after the *T*-jump. Other things being equal, a gas phase that cools more slowly will yield more product.

The rate of cooling of a hot gas is a complex physical phenomenon involving thermal conduction, radiation, convection, and other modes of mass transport. Whenever cooling by conduction to the cell walls is

Table 2-3 Effect of Cooling Rate on Conversion per Flash for $CHClF_2$ [10]

P (torr)	E_{abs} (kcal/mole)	Surface/Volume (cm^{-1})	CPF (%)
14.2	7.65	2.77	0.173
		7.00	0.22
13.0	18.0	2.77	15.2
		7.00	15.5
29.5	26.5	4.50	49.8
		7.00	45.2

important, the cooling rate may be increased by an increase in the surface-to-volume ratio of the irradiated cell. This criterion was applied to the decomposition of $CHClF_2$ [(2-11) and 2-12)], where CPF depends uniquely on E_{abs}. The experimental data are shown in Table 2-3. In this case, CPF is independent of surface-to-volume ratio within the limits of experimental error. Therefore it may be concluded that this reaction is not thermal, but photochemical. Had the reaction been thermal, the effects of changing the cooling rate should have been detected [10].

Effect of Sensitizers

A convincing proof for infrared photochemical reaction may often be obtained by the addition of an inert sensitizer. For example, in two separate experiments a mixture of 11.5 torr of CCl_2F_2 and 5.6 torr of SiF_4 was irradiated at 1031 cm^{-1} where only SiF_4 absorbs, as well as at 921 cm^{-1} where only CCl_2F_2 absorbs. SiF_4 was unreactive in both experiments. When the mixture was irradiated at 1031 cm^{-1}, less than 0.01% of the halocarbon was decomposed; however, when the frequency of the laser was changed to 921 cm^{-1}, CCl_2F_2 reacted to give $CClF_2CClF_2$, $CClF_3$, and minor products with a CPF of 0.24%. Interestingly enough, the energy absorbed from the pulse at 1031 cm^{-1} was twice the energy absorbed from the 921 cm^{-1} pulse [11]. The experiments prove that, when CCl_2F_2 absorbs directly, the activated molecules decompose in a time that is short compared to the time needed to exchange vibrational energy with SiF_4.

Chemical Evidence

When both components of an irradiated gaseous mixture are reactive but only one absorbs from the laser beam, the absorbing species may react photochemically while the other species reacts thermally, so that the two reactions are separated in time. Conversely, when the nature of the reaction products indicates that the two reactions are taking place at different times, reaction of the infrared-absorbing species must be photochemical.

For instance, when CCl_2F_2 is laser activated, it is clear from the nature of the dominant product, $CClF_2CClF_2$, that the primary step is the breaking of a carbon-chlorine bond [11]. When CBr_2F_2 is laser activated, the only detected product is $CBrF_2CBrF_2$, showing that the primary step is the breaking of a carbon-bromine bond. When an equimolar mixture of CCl_2F_2 and CBr_2F_2 is irradiated at a frequency at which only CCl_2F_2 absorbs, the dominant products are $CClF_2CClF_2$ and $CBrF_2CBrF_2$. There is no evidence for the formation of $CClF_2CBrF_2$, which surely would have been expected had the free-radical species $\cdot CClF_2$ and $\cdot CBrF_2$ been present simultaneously. Because CCl_2F_2 is the absorbing species, the experiment shows that its decomposition is photochemical, and that its reaction is practically finished before $\cdot CBrF_2$ is formed in appreciable quantities. CBr_2F_2 reacts during the *thermal* reaction period, while CCl_2F_2 does not, because of the markedly greater energy of the C-Cl relative to the C-Br bond.

Product Ratio

When both components of an irradiated gaseous mixture react thermally, and when the thermal rate constants have been obtained by independent methods, the ratio of the laser-induced reaction products may, in principle, be predicted. In practice, such prediction will be most reliable when the activation energies are nearly equal, so that the reaction temperatures need not be known precisely. If the gaseous mixture is thermally equilibrated before reaction takes place, the observed product ratio will agree with prediction. However, it is not clear that this necessary condition for thermal reaction is also a sufficient condition.

For instance, the activation energies for the thermal decompositions of ethyl acetate and isopropyl bromide in the gas phase [(2-13) and (2-14)] differ by only 0.2 kcal/mole, and the predicted product ratio for thermal reaction of a 3 : 1

$$CH_3\overset{\overset{\displaystyle O}{\|}}{C}OC_2H_5 \rightarrow CH_3CO_2H + CH_2{=}CH_2 \tag{2-13}$$

$$CH_3CHBrCH_3 \rightarrow CH_3CH{=}CH_2 + HBr \tag{2-14}$$

ester/bromide mixture is approximately 4. Under conditions where ethyl acetate is the infrared-absorbing species, the experimental product ratio is 4 ± 1 at relatively low doses, but changes so as to favor (2-13) at higher doses [15]. Clearly at the higher doses, part of the reaction of ethyl acetate takes place before thermal equilibration can occur.

Effect of Added Inert Gas on Yield

Addition of inert gas to an infrared absorbing reactant raises the heat-capacity of the gas phase and thus reduces the magnitude of the T-jump produced by a given E_{abs}. Because of the exponential dependence of the specific reaction rate on $1/T$, such addition would greatly depress the CPF of any thermal reaction. When thermal activation parameters are known, the magnitude of the thermal rate depression is roughly predictable. Inert gas addition also has a depressant effect on the CPF for a photochemical reaction, but that effect will be less marked. For instance, perfluorocyclobutane decomposes according to (2-15) both in thermolysis and on infrared irradiation at

$$cyclo\text{-}C_4F_8 \rightarrow 2\ C_2F_4 \tag{2-15}$$

949.5 cm^{-1} [16]. Addition of 20 torr of argon to 1 torr of C_4F_8 is predicted to reduce the CPF for thermal reaction by 10^{-7}–10^{-8}. The experimentally determined decrease, on the other hand, is only about one-sixth [16]. Thus the evidence rules against a thermal model.

Kinetic Spectroscopy

Experiments have been done in which a gas is exposed to an infrared laser pulse while the change in composition is being monitored as a

Fig. 2-4 Optical bench set-up in the authors' laboratory. (1) IR exit port of laser. (2) Holder for attenuators.Lens (if needed). (3) . For chemical studies, the IR beam then passes through appropriate aperture in plastic mask (5) and beam splitter (6), which reflects a small fraction to pyroelectric detector (7). Most of the beam continues through reaction cell (8) and is measured by disk calorimeter (10).

For kinetic spectroscopy, mirror (4) (under the dust cover) is placed to reflect IR beam along other optical bench; He-Ne laser (11) assists with optical alignment. The UV-VIS monitoring beam is produced by Xe-Hg arc (12),passes through the reaction cell (not visible on optical bench) at right angle to IR beam, is focused on monochrometers (13) and (14), and is measured by photomultipliers. Photo by John Galano (1977). Courtesy Brandeis University.

function of time with a continuous ultraviolet beam. By use of high-speed monitoring equipment it is thus possible to measure the speed of the infrared-induced primary process and compare it with known thermal rate constants. Some of the apparatus used in such experiments is shown in Fig. 2-4.

For instance, infrared flash photolysis of $CHClF_2$ leads to difluorocarbene, CF_2, in the primary step (2-11). The ultraviolet spectrum of CF_2 has been measured (although not at high temperatures) [17], allowing us to follow its formation and disappearance. Figure 2-5 shows the concentration of CF_2 (increase in concentration is down) as a function of time after the flash (whose power profile is shown by the sharp spike at the lower left of the figure) during the first 20 μs [18]. The events detailed by the figure are (1) the laser pulse of approximately 300 ns is followed by a reproducible delay of about 0.7 μs and (2) there is a rise of the CF_2 concentration with a half-time of approximately 3 μs. In the experiment shown, if the decomposition of $CHClF_2$ were thermal (with an appropriate *T*-jump to$\sim$1280°K), the half-time for CF_2 formation would have been greater than 100 μs. This is significantly greater than observation, and it may be concluded that the process is photochemical rather than thermal.

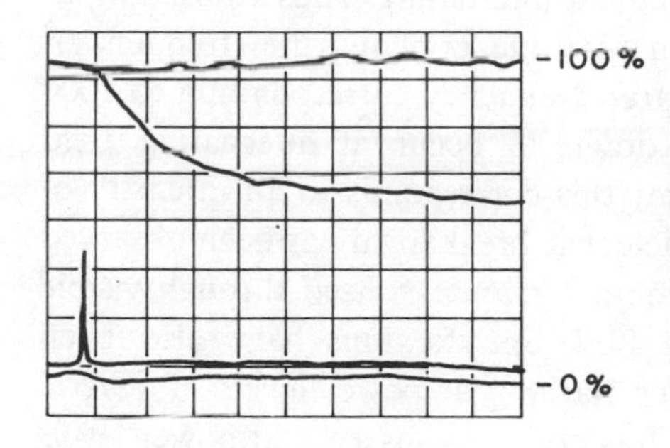

Fig. 2-5 Concentration of CF_2 vs time after flash irradiation of 12.5 torr $CHClF_2$ at 247.8 nm. 3.14 torr of CF_2 is produced. Horizontal scale 2 μs/div. Vertical scale, 100% transmission line at top, 0% line at bottom. Narrow spike at lower left is a photon drag signal which indicates the power profile of the laser flash. Courtesy of Dr. C. M. Lonzetta.

NATURE OF LASER-INDUCED REACTIONS

The primary reaction steps of megawatt laser-induced reactions occur in the electronic ground state, and reaction products are qualitatively the same as those produced in conventional high-temperature chemistry. There may be quantitative differences, however, owing to the high "effective temperatures" that can be reached, which favor products formed with a high activation energy. Although most known primary reaction steps are decompositions, cases are known in which the primary reactions are isomerization or rearrangement, and there are also likely examples in which they are bimolecular addition or substitution. Further details are presented in Chapter 5.

Secondary reactions of the primary decomposition products may lead to electronically excited species. In such cases, one may observe visible emission. For example, the infrared laser-induced decomposition of SF_6 in the presence of H_2 or of organic hydrogen compounds is accompanied by emission of a bright flash of light whose color depends on the nature of the hydrogen compound [8]. One of us (D.F.D) has observed a bluish-white emission when $CClF_2CClF_2$ is excited in the presence of H_2 and some hydrocarbons. Light emission accompanying infrared laser excitation becomes fairly common at gigawatt power levels [2, 19, 20].

Finally, it should be noted that reactions leading to ions, similar to those commonly observed in microwave discharges, are not expected with megawatt infrared laser radiation.* The efficiency with which energy is transferred from a high-frequency alternating electromagnetic field (in this case, the laser beam) to electrons in a gas phase varies inversely with the square of the frequency [21]. At microwave frequencies, ion formation by dielectric breakdown is known to occur at alternating fields whose amplitudes are of the order of 10^2 V/cm [21]. On applying the $(\text{frequency})^2$ relationship and changing to an infrared frequency corresponding to 1000 cm^{-1}, one expects dielectric breakdown to occur at alternating field amplitudes of the order of 10^6 V/cm; this corresponds to an intensity of the order of 1 GW/cm^2.** In fact, dielectric breakdown has been observed at power levels of this magnitude, where it manifests itself through visible luminescence followed by sparking [22]. Specific ions have also been identified by mass spectrometry after flashing at power levels of several GW/cm^2 [23]. Dielectric breakdown has not been observed at power levels well below the theoretical threshold. For instance, in our own experiments we see no sparks, nor do we find products whose formation requires ionic reaction mechanisms.

*Secondary reactions involving electron transfer, and thus yielding ions, are possible in principle, but are expected only in unusual special cases owing to the high Born charging energy of molecules in the gas phase.

**Root-mean-square intensity I (J/cm^2) is related to amplitude E_0 (V/cm) of the alternating electric vector associated with the coherent radiation by $I = E_0^2/754$.

Chapter 3

Some Photophysics

Absorption of radiant energy from high-power coherent infrared sources is an active area of research in molecular physics. It is fair to say that the understanding of this phenomenon is still incomplete. At least some of the uncertainty is due to the paucity of data. Yet the absorption of radiation is so fundamental to a consideration of infrared laser chemistry that we cannot avoid a brief pragmatic discussion. In the following we emphasize facts, and we introduce models only to the extent that they appear to be useful.

COHERENCE

Any discussion of absorption of radiation from laser sources must begin with a consideration of coherence. As is well known, an ordinary incandescent or fluorescent source is *chaotic* because the individual atomic events resulting in emission are independent of one another. Thus there are continual random fluctuations in amplitude and phase of the emitted light.

By contrast, a source of stimulated emission is *coherent*: the radiation amplitude at a given point remains a simple sine or cosine function of time for a finite length of time, whose average is called *coherence time*. Similarly, the radiation amplitude at a given time is a simple sine or cosine function of distance for a finite distance along the direction of propagation. There is also a spatial coherence transverse to the direction of propagation, owing to the random motion of molecules before they undergo stimulated emission.

So far, we have talked about coherence of *radiation*. For a molecular oscillator (the absorbing molecule) it is useful to talk about coherence of *vibration* as well. For an isolated molecule, the coherence of vibration persists indefinitely. However, collisions with other molecules disrupt that coherence, even when such collisions are elastic. A molecular collision is analogous to a half-cycle of disturbing vibration. When the original molecule emerges from the collision, the phase of the motion is almost certainly changed.

When coherent radiation is absorbed by a coherently vibrating molecule, the absorption process is also said to be coherent. When either radiation or vibration is incoherent on the relevant time scale, the absorption process is said to be incoherent. To see why the distinction is important, consider a classical oscillator (the molecule) interacting with a resonant driving force (the radiation). The magnitude of the power transfer depends on the phase relationship between the two motions; the more closely they match, the more efficient the transfer.

In order to discuss this distinction, let us consider some simplified sets of events. In the case of laser sources for which radiation is coherent, the coherence of the absorption process may be decided by either of two criteria. According to the *spatial* criterion, absorption is coherent if the mean free path for translational motion exceeds the radius of the laser beam; according to the *temporal* criterion, one assumes that there is a characteristic time for the absorption of a photon by a molecule. If the time between collisions is short compared to it, that is, if there are a number of collisions between successive absorption-emission events, the absorption is said to be incoherent with respect to time. The characteristic time varies inversely as the intensity of the radiation, and the time between collisions varies inversely as the pressure. The first may be estimated in terms of the extinction coefficient, according to (3-1).

$$\left\{\begin{matrix}\text{characteristic time for}\\ \text{successive photon events}\end{matrix}\right\} \cong \left\{\begin{matrix}\text{mean turnover time}\\ \text{of the molecule in}\\ \text{the vibrational}\\ \text{ground state}\end{matrix}\right\} = \frac{h\nu/kT}{2300\,\varepsilon_A I} \qquad (3\text{-}1)$$

On the right-hand side of (3-1), I is expressed in W/cm^2, ε_A in $torr^{-1}/cm$, and $h\nu/kT$ in consistent units. In megawatt infrared laser chemistry, this characteristic time has values in the order of 10^{-8}–10^{-7} s. Mean time between collisions will be $<3\times10^{-8}$ s when the pressure is greater than approximately 2 torr. In the majority of experiments under discussion, absorption is incoherent by both criteria.

DOSE AND INTENSITY MEASUREMENT

In reading the literature dealing with pulsed lasers, one encounters the following terminology.

	energy per pulse and *power*	refer to the entire laser beam
I	*beam intensity* (or simply *intensity*)	radiant power per square centimeter (or radiant flux per square centimeter)
D	*dose*	radiant energy per square centimeter, per pulse, as defined in (3-2); also called *energy fluence*
I_D	*dose-average intensity*	average intensity at which the dose is delivered; defined in (3-3)
τ	*effective pulse duration*	time necessary for the average intensity to be in flux to deliver the measured dose; defined in (3-4)
	power profile	a graph showing power as a function of time during the laser pulse
FWHM	*full width at half-maximum*	a measure of laser pulse width

$$D=\int I dt; \quad \text{integration over one pulse} \tag{3-2}$$

$$I_D=\frac{\int I dD}{D}=\frac{\int I^2 dt}{\int I dt}; \quad \text{integration over one pulse} \tag{3-3}$$

$$\tau=\frac{D}{I_D} \tag{3-4}$$

The concepts of power profile, dose-average intensity, and effective pulse duration are depicted in Fig. 3-1.

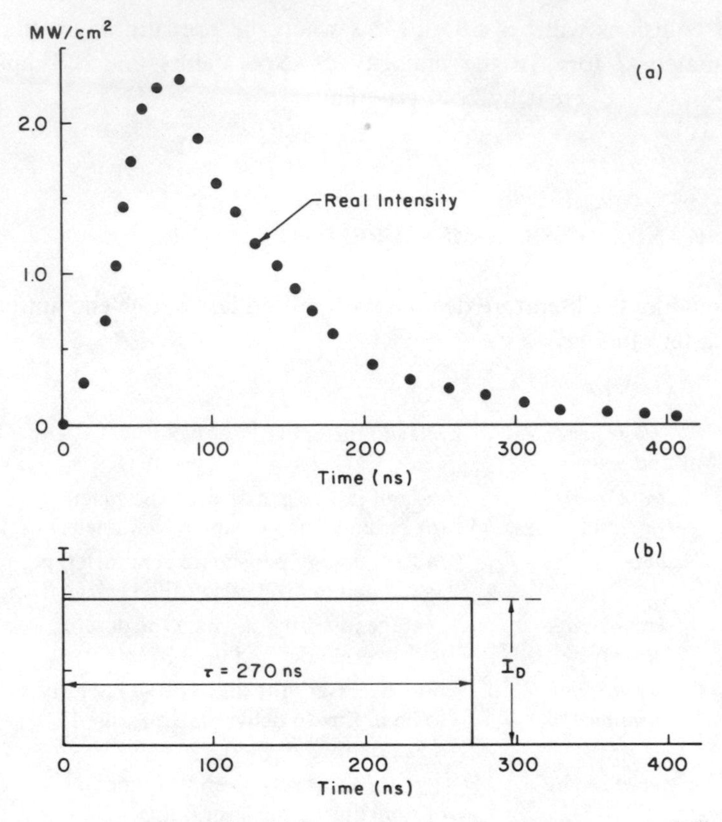

Fig. 3-1 Laser intensity profiles. (*a*) Outline of real profile; fine structure, including intensity spikes, not shown. (*b*) Equivalent step-function.

Dose

One measures the amount of radiant energy passing through a known area. The problems involved in getting accurate data are similar to those in conventional calorimetry. The radiant energy must be fully absorbed, producing a temperature rise. The resulting electrical signal must be measured and calibrated in terms of energy. There are several different techniques available; two of the more important ones are discussed and compared.

The *pyroelectric detector* consists of a ferroelectric Li/Ta disk, 5 cm in diameter, mounted on a heat sink [24]. This disk, held perpendicular to the laser beam, is flat-black coated with an antireflection paint which contains sintered glass to increase internal reflections, thus enhancing black-body characteristics. The ferroelectric disk has the electrical properties of an array of partially ordered dipoles, and thus there is a potential difference across it. When radiant energy is absorbed, the temperature of the disk rises, causing a decrease in the dipole ordering and thus of the observed potential difference. As heat subsequently flows from the disk into the heat sink, the original potential difference is restored. Typical outputs are in the range of 1.5 V/J; detectors usually come calibrated. An advantage is the short recovery half-time, approximately 20 ms. Disadvantages include lack of high accuracy and, after some use, lack of uniformity across the detector surface [11]. The device is most useful when energy must be measured at relatively high repetition rates.

The *disk calorimeter* consists of a radiation-absorbing disk connected by way of a heat-conducting pipe to a heat sink. The temperature difference between disk and heat sink is measured by means of a thermopile. A schematic diagram is shown in Fig. 3-2. The device was developed at the U.S. Bureau of Standards and is capable of high accuracy [25]. The temperature rise and eventual return to room conditions (with a half-time of ~ 7 s) are monitored by the thermopile and produce thermal emf's of the order of 1 mV or less. The integral of the voltage over time is proportional to the energy falling on the disk.

A heating wire is mounted on the disk to permit direct electrical calibration of the instrument. The thermal effect of the laser pulse can be simulated by discharging a capacitor at a known voltage through this heating wire [10]. The advantages are (1) the instrument can be calibrated at an energy level close to that of the laser output, and (2) calibrations can bracket in time the measurements of radiant energy, without disturbing any optical bench settings. The disadvantages are (1) a substantial amount

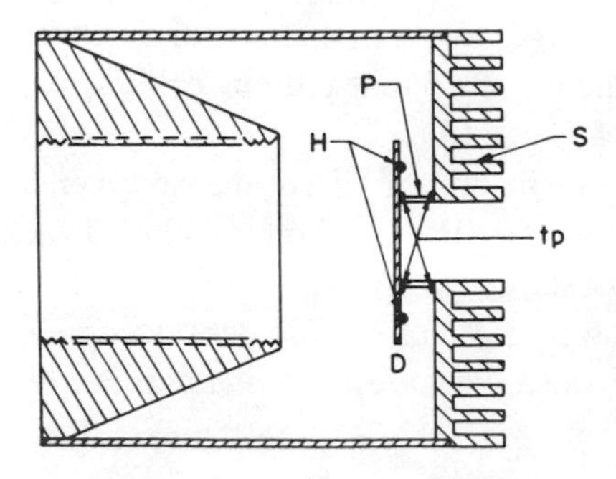

Fig. 3-2 Cross-section of disk calorimeter. D = absorbing disk; S = heat sink with cooling fins; P = thin heat-conducting pipe; H = heater wire, for energy calibration; tp = thermopile. For details, see [25].

of instrumentation is needed for calibration and use, and (2) the recovery time for this instrument is relatively long; thus every pulse, at the usual pulse rate, cannot be monitored. As seen below, this last objection is not necessarily serious.

Amount of Absorbed Energy

Two methods are in common use: direct measurement with optoacoustic detection; and indirect measurement, based on the decrease in transmitted dose.

The *optoacoustic detector* is based on an effect first reported by Alexander Graham Bell almost a hundred years ago [26]. He described experiments in which sunlight, chopped at audio frequencies, was allowed to fall on the open end of a cylinder; when a stethoscopelike tube was attached to the other end of the cylinder, a "musical" note was heard. A recent account [27] describes a detector that uses the optoacoustic effect to measure radiant absorption in spectroscopy. The detector is useful over a wide range of wavelengths, from the infrared into the ultraviolet.

When used as an energy detector for pulsed infrared laser experiments, the detector takes the shape of a condenser microphone placed *inside* the reaction cell [28]. When the gas is irradiated by the laser pulse, the absorbed energy is present initially as vibrational energy. As that energy is transferred into translational modes, the increase of the translational temperature is accompanied by a pressure increase and produces an acoustical pulse of approximately 1 ms duration. This signal, proportional to the energy absorbed, is detected by the microphone. The technique has been used to measure a wide range of energies absorbed by SF_6 from a TEA CO_2 laser pulse [29]. The advantage of this technique is that energy absorbed is measured directly. Disadvantages are that the signal is noisy owing to acoustical standing waves in the cell, and that tedious calibration is necessary to obtain absolute values of E_{abs} [30].

When E_{abs} is at least 10% of the incident dose, indirect measurement, by *monitoring the decrease in transmission*, is the method of choice. A convenient experimental arrangement is shown in Figs. 3-3 and 3-4 [10]. The beam splitter diverts a small fraction ($\sim$10%) of the incident dose to a pyroelectric detector to permit relative dose measurement of consecutive

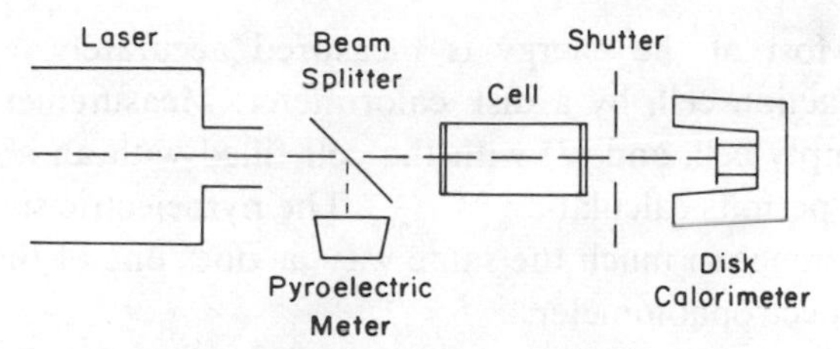

Fig. 3-3 Optical path for indirect measurement of absorbed energy. See Fig. 3-4 for photograph of laboratory arrangement.

Fig. 3-4 Dr. K. J. Olszyna with infrared dose-measuring apparatus. Component-layout is essentially the same as in Fig. 3-3. The laser beam exits from the Faraday cage in the back and is reflected by a plane mirror into the measuring apparatus. Photograph by Ralph Norman (1976). Courtesy Brandeis University.

laser pulses. Most of the energy is measured accurately, after passage through the reaction cell, by a disk calorimeter. Measurements are made (1) with the empty cell, and (2) with the cell filled with an absorbing gas. The difference permits calculation of E_{abs}. The pyroelectric signal serves as an internal reference in much the same way as does one of the beams in a double-beam spectrophotometer.

Intensity

A convenient instrument for measuring intensity and power profiles for pulsed lasers is the photon drag detector. The detector element of this instrument is a disk of gallium-doped germanium sufficiently transparent in the infrared to allow for a substantial fraction of the radiation to penetrate beyond the surface. This radiation interacts with the free carriers in the disk, causing a voltage change of approximately 125 mV per MW/cm^2. This signal, having an impedance of 50–100 ohms, is captured on a fast storage oscilloscope. The rise time of the signal is on the order of 1 ns, so that pulse times of hundreds of nanoseconds commonly associated with pulsed CO_2 TEA lasers are aptly measured with this detector. A typical laser power profile, though on a compressed scale, is included in Fig. 2-4.

TYPICAL RESULTS

Figure 3-5 shows, schematically, three types of absorption behavior, all of which have been observed [29]. Saturative behavior (*c*) is perhaps the most common, although genuine saturation, such that dE_{abs}/dD approaches zero and the sample practically becomes "bleached," is fortunately rare.

When examined in detail, the absorption behavior of gases at megawatt laser power levels is quite complex, even for a single substance and a single absorption band. For instance, for the 949 cm^{-1} out-of-plane bending mode of ethylene, absorption has been examined on both sides of the band maximum [31]. On the high-frequency side, at 951.2 cm^{-1}, absorption is saturative; on the low frequency side, at 949.5 and 947.7

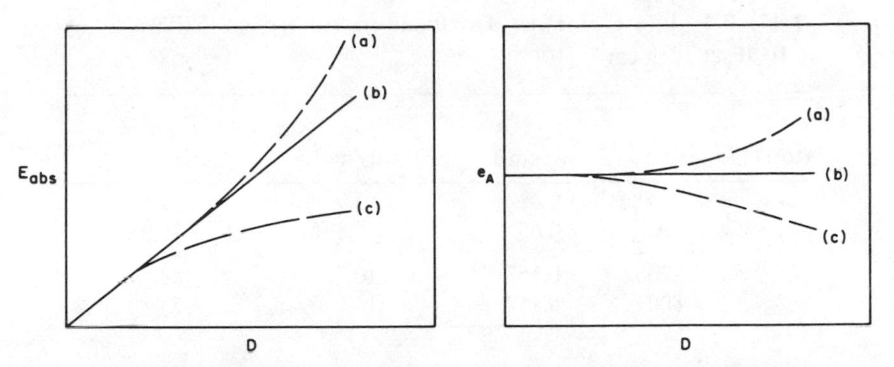

Fig. 3-5 Absorption behavior vs dose: (*a*) accelerative, (*b*) linear, (*c*) saturative.

cm^{-1}, the extinction coefficient e_A is independent of dose, although it varies with frequency. Saturative behavior in general is more likely to occur on the high-frequency side than on the low-frequency side near the band maximum; other things being equal, e_A goes through a maximum at a somewhat lower frequency than does ε_A. This behavior may be accounted for in terms of anharmonic frequency shifts, as explained later in this chapter.

It is clear from the above that the relation of e_A to the spectrophotometric extinction coefficient ε_A is complicated. Knowledge of the relative absorption efficiency, e_A/ε_A, for one band envelope does not allow us to predict very well the relative absorption efficiency for another band envelope of the same compound. Indeed, the band which absorbs more efficiently by spectrophotometry may absorb more poorly from a pulsed megawatt laser beam. For instance, for CCl_2F_2 at 12 torr, the relative absorption efficiency is 0.19 at 921 cm^{-1} and 0.52 at 1088 cm^{-1}. The absolute value of e_A is greater at 1088 cm^{-1}, while that of ε_A is greater at 921 cm^{-1} [11].

Pulse Duration

In contrast to the relation of CPF to E_{abs} which is independent of pulse duration (for example, Table 2-2), the relation of E_{abs} to dose depends on the amount of time during which the dose is delivered. Because the absorption of radiation is a rate process, one might expect that e_A increases with τ. Quantitative data are scarce. Some results for $CHClF_2$ are

Table 3-1 Effect of Pulse Duration on Absorption Efficiency of $CHClF_2$at 1088 cm^{-1} [10]

P (torr)	τ (ns)	D (J/cm^2)	E_{abs} (kcal/mole)	E_{abs}/D (efficiency)
1.6	270	0.41	3.9	9.5
	800	0.65	18.8	28.9
50	270	0.157	10.7	68
	800	0.157	11.5	73

shown in Table 3-1. At 1.6 torr, increasing the pulse duration from 270 ns to 800 ns increases the absorption efficiency by a factor of 3. At 50 torr, a qualitatively similar effect is seen, but it is much less marked [10].

Absorber Pressure

The indication from the effect of pulse duration, that molecular collisions play a part in infrared absorption, is confirmed by the effect of pressure. We again examine data for neat $CHClF_2$ because of the relatively simple relationships which obtain when $6 < E_{abs} < 25$ kcal/mole [10]. In this range, and at constant pressure, E_{abs} is practically proportional to dose; however, the slope E_{abs}/D varies with the pressure. An empirical relationship which reproduces the data is given in (3-5).

$$\frac{E_{abs}}{D} = \frac{92.8\, P_A}{P_A + 22.1} + 4.5 \tag{3-5}$$

Here E_{abs} is in kcal/mole, D in J/cm^2 at 1088 cm^{-1} P_A in torr of $CHClF_2$ and τ is 270 ns. Note that E_{abs}/D approaches an asymptotic upper limit as P_A becomes large. Note also that at low pressure, E_{abs}/D does not go to zero. The fact that E_{abs}/D remains finite even at very low pressures is crucial in current research on infrared laser isotope separation of uranium.

In more complicated cases, where absorption behavior is accelerative or saturative (Fig. 3-5), a more general empirical relationship must be used to accomodate both pressure and intensity dependence. Equation 3-6 has been proposed and represents data for SiF_4 over a wide range of variables [32].

$$\frac{E_{abs}}{D} = \frac{A}{(1+B/P)(1+CI_D)} + \alpha \tag{3-6}$$

A, B, C, and α are adjustable parameters. A, B, and α are positive quantities; C will be positive, zero, or negative depending on whether the absorption behavior is saturative, "linear," or accelerative.

Effect of Nonabsorbing Gases

The importance of molecular collisions in infrared absorption is further emphasized by experiments in which absorption takes place in the presence of nonabsorbing added gas. Without significant exception, such additions have led to an increase in E_{abs}/D. The effect has been studied in some detail for $CHClF_2$ in the presence of He, N_2, and SiF_4 [10]. Absorption continues to take place according to (3-5) except that an effective pressure, P_{eff}, must be used in place of P_A, the absorber pressure. The effective pressure is given by (3-7).

$$P_{eff} = P_A + KP_X \tag{3-7}$$

P_X denotes the partial pressure of the added nonabsorbing gas and K is a parameter measuring the relative efficiency of collisions with X (the added gas) in promoting infrared absorption by A.

Values for K obtained for each gas are He, 0.21; N_2, 0.38; SiF_4, 0.7. The quality of agreement with (3-5) and (3-7) is illustrated in Fig. 3-6. Note that K increases with the molecular weight of X. Such an increasing trend is to be expected if the amount of energy transferred in the collision is small [33].

Analogous phenomena have been observed for infrared absorption by SiF_4. However, the detailed relationships are more complicated because infrared absorption is saturative. Typical results are shown in Fig. 3-7, in which various amounts of cyclopropane are added to SiF_4 [8]. For neat SiF_4 at 19 torr, absorption becomes markedly saturative when D is greater

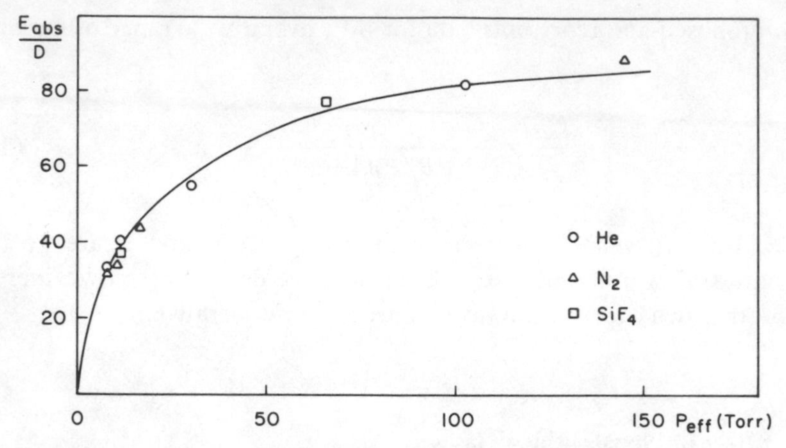

Fig. 3-6 Effect of nonabsorbing gases on E_{abs}/D for $CHClF_2$ at 1088 cm^{-1}. Based on data in [10].

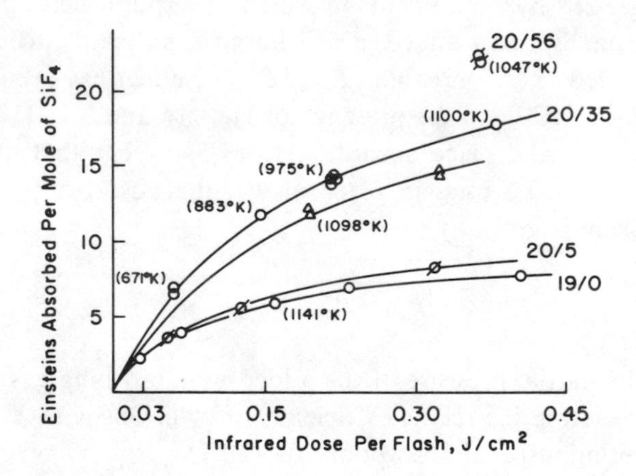

Fig. 3-7 Absorption behavior of SiF_4 at 1025 cm^{-1} in the presence of cyclopropane, C_3H_6. Numbers on the right, such as 20/56, denote torr SiF_4/torr C_3H_6. Values of °K adjacent to the experimental points denote the hypothetical temperatures that would be reached if E_{abs} were converted adiabatically into random thermal energy. Based on data in [8].

than 0.15 J/cm^2. The tendency toward saturation is greatly reduced by 35 torr or more of cyclopropane. The latter gas by itself is practically nonabsorbing at the given frequency. The "temperatures" shown in Fig. 3-7 are hypothetical, indicating how hot the gas would become if E_{abs} were converted into random thermal energy. Most of the values are above 1000 °K.

THEORETICAL MODELS

Harmonic Oscillator

The concept of the harmonic oscillator is too simple to be a real model for polyatomic molecules; it is also too simple for the description of diatomic molecules in all but the lowest vibrational states. Yet, when considering the complexities of incoherent absorption of infrared laser radiation, the model is worthy of our attention, because (in common with any good fiction) it brings to light relevant issues and helps us to understand them.

To consider the absorption of radiation by an ensemble of harmonic oscillators, we take one mole of oscillators, all at the same frequency. Let $f_0, f_1, f_2, \ldots$ denote the mole fractions of oscillators in quantum state $v=0, 1, 2\ldots$, respectively. We shall make no assumption as to statistical equilibrium, but will assume the existence of conditions under which absorption is incoherent.* In the presence of a radiation field of intensity I, energy is absorbed in accordance with selection rules for a harmonic oscillator ($\Delta v = \pm 1$; Fig. 3-8). For incoherent absorption, transition probabilities $\alpha_{k,k+1}$ and $\alpha_{k+1,k}$ are given by (3-8) and (3-9).

$$\alpha_{01} = \alpha_{10} \equiv \alpha \tag{3-8}$$

$$\alpha_{k,k+1} = \alpha_{k+1,k} = (k+1)\alpha \tag{3-9}$$

The differential equation for E_{abs} as a function of time is therefore (3-10)

$$\begin{aligned} \frac{dE_{\text{abs}}}{dt} &= I[\alpha f_0 + (2\alpha - \alpha) f_1 + (3\alpha - 2\alpha) f_2 + \ldots] \\ &= I\alpha \sum_{k=0}^{\infty} f_k = I\alpha \end{aligned} \tag{3-10}$$

Integration leads to (3-11)

$$E_{\text{abs}} = \alpha \int I\,dt = \alpha D \tag{3-11}$$

The following features of the model are worth noting: (*a*) E_{abs} is a function of the radiant dose and does not depend on the detailed manner

*Coherent absorption is considered in the final section of this chapter.

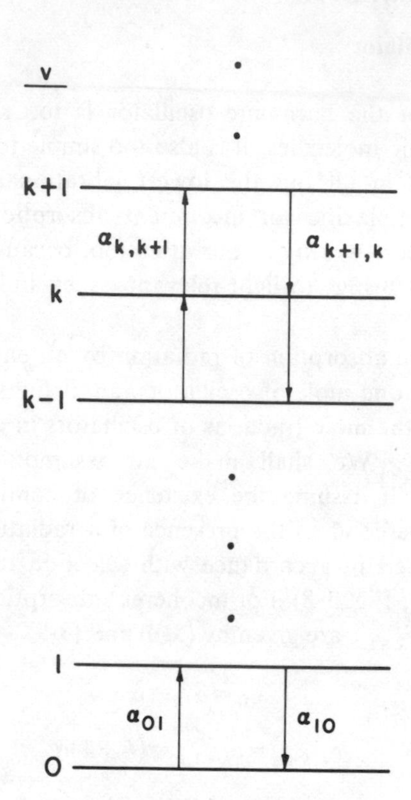

Fig. 3-8 Transition probabilities for the harmonic oscillator.

$I(t)$ in which the dose is delivered by the laser, (*b*) E_{abs} is proportional to dose, the constant of proportionality being the 0→1 transition probability α, (*c*) Deviations of the populations $f_1, f_2, f_3 \ldots$ from statistical equilibrium have no effect on the absorption of radiant energy.

In addition to these features, which are implied by (3-10) and (3-11), two further features are stated without proof. (*d*) If the initial energy distribution of harmonic oscillators is a Boltzmann distribution, then the transition probabilities [given in (3-9)] work in such a way that the final distribution, after absorption, is also a Boltzmann distribution, characterized by a higher temperature. (*e*) If the initial energy distribution is *not* a Boltzmann distribution, molecular collisions of the type shown in (3-12) will be extraordinarily effective in establishing a Boltzmann distribution.

$$(v=k)+(v=l)\rightarrow(v=k-1)+(v=l+1) \tag{3-12}$$

Comparing absorption behavior of real molecules with that of an ensemble of harmonic oscillators, we find that saturative and accelerative types of behavior are not accounted for. If we ascribe the deviations to anharmonic transition probabilities and assume that these effects increase monotonically up the vibrational ladder, we obtain the following propensity rules:

For saturative absorption:

$$\alpha_{k,k+1}<\frac{(k+1)}{k}\alpha_{k-1,k} \tag{3-13}$$

For accelerative absorption:

$$\alpha_{k,k+1}>\frac{(k+1)}{k}\alpha_{k-1,k} \tag{3-14}$$

P/Q/R/Continuum Model

More realistic models allow for the facts that molecular oscillators are anharmonic, and that infrared absorption bands of polyatomic molecules show broadening and structure due to coupling of vibration and rotation. The spacing of anharmonic vibrational levels decreases as the molecule moves up the ladder, while the laser frequency is fixed. Thus if the laser is tuned to the maximum for the absorption for the $0\rightarrow1$ transition, there will be a progressive frequency mismatch as the transitions involve higher and higher levels. This mismatch can be compensated for by taking advantage of (1) the rotational band structure and (2) the rapid increase in the density of levels with increasing energy.

The essence of the P/Q/R/Continuum model is shown in Fig. 3-9 [2,34]. It is assumed that vibrational levels are essentially discrete up to $v\simeq3$, but that for $v\geqslant4$ ($\geqslant12$ kcal/mole) the density of vibrational levels for the molecule is already so high that the vibrational manifold is a quasi-continuum. The high level density at high energies results from the

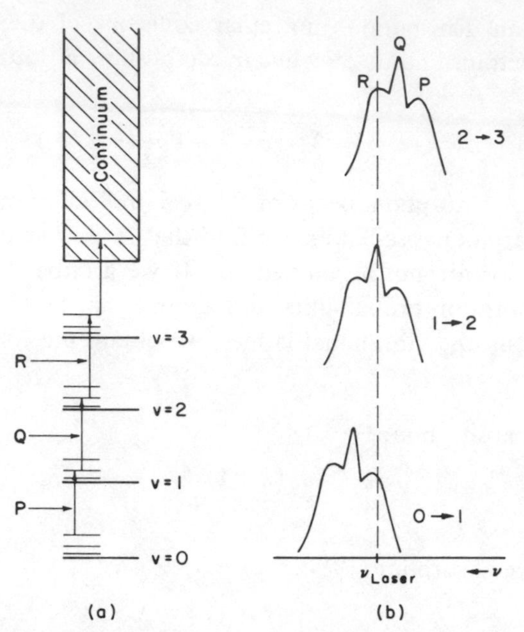

Fig. 3-9 Features of P/Q/R/Continuum model. (*a*) Exhibits passage of mode through P, Q, and R branches on into continuum. (*b*) Shows anharmonic frequency shifts as molecule is excited to higher and higher energy states.

fact that there are many ways in which the given amount of energy can be distributed among the various modes. The practical implication of this model is that, once excitation to $v \geqslant 4$ has been accomplished, further excitation takes place in a quasi-continuum and is no longer hampered by quantum restrictions. The secret to achieving efficient absorption is therefore to overcome the bottleneck between $v=0$ and $v=4$.

This is where the "P/Q/R" part of the P/Q/R/Continuum model enters the picture. For example, the infrared absorption band shown in Fig. 3-9*b* features distinct branches labeled P, Q, and R which correspond to rotational changes with $\Delta J = -1, 0, +1$ occurring along with the more energetic vibrational transitions. If the laser is tuned to the P branch of the $0 \rightarrow 1$ vibrational transition, then the anharmonic frequency shifts in progressively higher excitation steps will cause the laser frequency to move into the regions of the Q and R branches. For best results, it has been suggested [2] that the laser frequency should coincide with the Q branch of the $1 \rightarrow 2$ transition. This will happen when the laser frequency is equal to ν_1-ν_0, the difference between the first overtone and the fundamental.

As indicated in Fig. 3-9*a*, only *some* P-branch molecules are in resonance with the laser frequency and thus can be excited to an upper level. Of those in the upper level, only *some* Q-branch molecules* are in resonance with the laser frequency and can be excited. And so on up the ladder, except that above $v=4$ it does not seem to matter. In order to have efficient excitation of *all* molecules, it is evidently necessary that those molecules which are not in an excitable rotational state be moved into one, so that they may absorb the radiation. This can be accomplished by inelastic collisions with other molecules.

Besides transferring molecules between rotational states, collisions also broaden the fine structure of vibrational absorption bands and thus facilitate matching to the laser frequency. The band fine structure is broadened also by the radiation field itself. However, this effect does not become highly important until gigawatt power levels are reached. [2, 34, 35]

The preceding model is in agreement with two aspects of the experimental facts. First, it reproduces the fact that laser excitation is usually most efficient at a somewhat lower frequency than that of the spectrophotometric absorption maximum. Second, it is qualitatively consistent with the observed pressure effect discussed in the preceding section.

Coherent Versus Incoherent Absorption

At gas pressures that are high enough so that infrared absorption is incoherent, the resulting molecular energy distributions are broad, and there is considerable reaction even when E_{abs} is several times smaller than the activation energy for reaction. Representative data are shown in Table 2-1 and are discussed in that connection.

On the other hand, when infrared absorption takes place coherently (from a coherent source under collision-free conditions), there is practically no reaction until E_{abs} is almost as great as the activation energy [2]. The onset of reaction then is quite sharp and may be described approximately as a reaction threshold.

These seemingly contradictory facts can find a logical explanation in terms of the harmonic oscillator model. When an ensemble of harmonic oscillators absorbs radiant energy *in*coherently according to the transition probabilities [see (3-9)], the resulting Boltzmann distribution is a *broad*

*Unless the Q branch is genuinely a sharp "spike".

distribution. On the other hand, when the oscillators absorb coherently, the resulting distribution is the relatively narrow Poisson distribution. In probability theory, a Poisson distribution results when the probabilities for consecutive events are independent of one another, as in the familiar example that r balls are placed at random into n baskets. When r and n are both large numbers, the Poisson distribution function gives the probability f_k that a given basket chosen at random will hold k balls, where $k = 0, 1, 2, \ldots$. According to quantum theory of radiation, an ensemble of independent oscillators, unperturbed by collisions and interacting with a coherent radiation field, is expected to gain energy according to Poisson statistics [36, 101].

Let $\bar{v}$ denote the mean excitation number [as in (3-15)] for the ensemble. For a Poisson distribution, the probability f_k that a given oscillator will have k quanta of excitation energy is given by (3-16), while for a Boltzmann distribution the analogous probability is (3-17).

1. Mean excitation number

$$\bar{v} = \sum_{k=0}^{\infty} k f_k \tag{3-15}$$

2. Excitation probability in a Poisson distribution

$$f_k = (\bar{v})^k \cdot \frac{e^{-\bar{v}}}{k!} \tag{3-16}$$

3. Excitation probability in a Boltzmann distribution

$$f_k = \frac{(\bar{v}-1)^k}{\bar{v}^{k+1}} \tag{3-17}$$

The two distributions are compared in Fig. 3-10 for the case that $\bar{v} = 10$. At an oscillator frequency corresponding to 1000 cm^{-1}, $\bar{v} = 10$ corresponds to 28.6 kcal/mole of excitation energy.

As shown in Fig. 3-10, in the Boltzmann distribution f_k decreases monotonically with increasing k, the rate of decrease being relatively gentle. Thus, when $k = 30$, $f_k = 0.0042$, and the total fraction of molecules

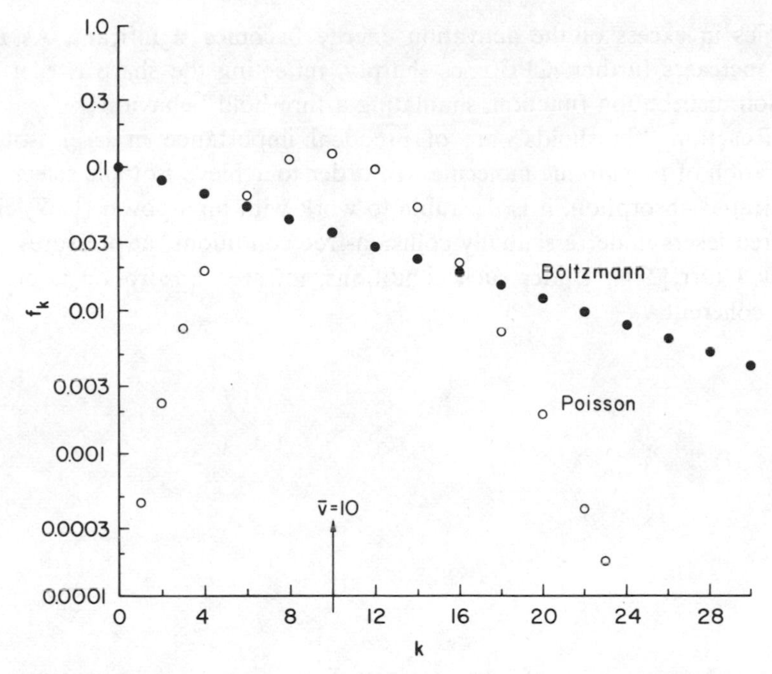

Fig. 3-10 Comparison of Boltzmann and Poisson distributions. Mean excitation number $\bar{v} = 10$.

with $k \geqslant 30$ [which equals $\bar{v} \cdot f_k$ for any value of k, as a consequence of (3-17)] is 0.042. This is a remarkably large fraction, especially when one considers that the vibrational energy for this fraction is at least three times the mean for the ensemble.

By comparison, in the Poisson distribution, f_k goes through a maximum when $k = \bar{v}$ and decreases rapidly as k becomes large. Thus, when $k = 30$, f_k is only 1.7×10^{-7}, and the total fraction with $k \geqslant 30$ is only 2.5×10^{-7}. In fact, a Poisson distribution with $\bar{v}$ greater than about 6 is quite similar to the familiar Gaussian "normal curve" with mean equal to $\bar{v}$ and standard deviation equal to $\bar{v}^{1/2}$.

An ensemble of reactive molecules whose excitation leads to a Poisson distribution will display an approximate reaction threshold. Because the fraction of molecules with energies greatly in excess of the mean is relatively small, reaction will be quite negligible as long as E_{abs} is considerably less than the activation energy for reaction. Reaction becomes observable only as E_{abs} approaches E_{act} and the fraction of molecules with energies in excess of the activation energy becomes significant. As E_{abs} then increases further, CPF rises sharply, reflecting the sharp rise of the Poisson distribution function, simulating a threshold behavior.

Reaction "thresholds" are of practical importance in laser isotope separation of polyatomic molecules. In order to achieve isotopic selectivity of infrared absorption, it is desirable to work with high-power (GW/cm^2) infrared lasers under essentially collision-free conditions, at pressures well below 1 torr [2–4]. Under such conditions, infrared absorption is practically coherent.

Chapter 4

Molecular Energy-Flow Patterns

The term "mechanism" in infrared laser chemistry has a broader meaning than it does in classical mechanistic chemistry. It includes not only plausible, detailed reaction steps connecting both sides of a stoichiometric equation but, as an earlier priority, concerns itself with the energy-flow patterns from injection of energy into the mode of excitation to propulsion of the molecule through the reaction channel of the primary process. In this chapter we consider such patterns. It will be assumed throughout that the gas pressure is high enough so that infrared absorption is incoherent.

Knowledge of the energy-flow patterns is important because many reactions take place before the absorbed energy has become random thermal energy. An infrared laser-induced reaction which takes place from a nonequilibrium energy distribution differs in two ways from a conventional thermal reaction. The primary step (bond scission, elimination, etc...) takes place in a very short time, less time than it takes for the energy to flow into rotational and translational modes (the heat bath), so that extraordinarily high concentrations of labile species are produced in a short time; the production rates are so high that surprisingly large amounts accumulate before they can react. For example, in the decomposition of $CHClF_2$, the formation of $:CF_2$ was shown to occur with a half-time of 3 μs (Fig. 2-5), during which time 3 torr of the carbene was produced.

The other difference is that the fraction of molecules whose energy exceeds the activation threshold is not the same, and may be greater, than it would be if the same amount of energy were distributed thermally. This concept is sufficiently important to be discussed more fully.

QUALITATIVE DISCUSSION

Because the laser radiation is monochromatic, infrared excitation must take place in equal energy increments, $h\nu$. If there were no collisions to redistribute the energy, the molecules would be constrained to exist in those levels, called *levels of excitation*, that can be reached through absorption of the given energy quantum. According to the P/Q/R/Continuum model, the levels of excitation are sharp and equidistant, with wide gaps between them (see Fig. 4-1*a*). Conversion of the excitation energy into random thermal energy is a complex process and includes these gaps being filled (Fig. 4-1*b*), as well as energy being transferred into rotation and translation. These events occur as a result of collisions. Even though

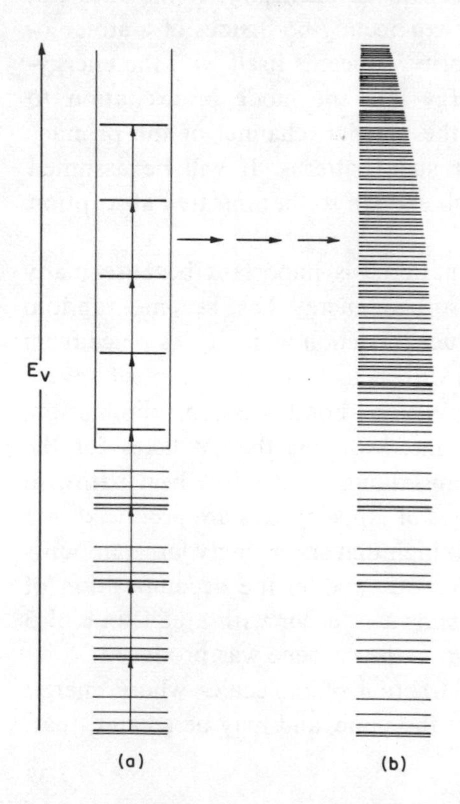

Fig. 4-1 Molecular energy distribution according to P/Q/R/continuum model of excitation. (*a*) Hypothetical distribution limited to levels specified by absorbed quanta. (*b*) Subsequent random thermal redistribution over states in continuum.

Table 4-1 Effect of Delocalization of Vibrational Energy (E_{abs}) over a Number (s) of 1000 cm^{-1} Harmonic Oscillator Modes

E_{abs}	E_{act}	Fraction of Molecules with $E \geqslant E_{act}$ when		
(kcal/mole)	(kcal/mole)	$s=1$	$s=2$	$s=5$
10	40	2.96×10^{-2}	9.8×10^{-3}	1.5×10^{-3}
10	80	8.8×10^{-4}	3.4×10^{-5}	7.0×10^{-8}
20	40	1.54×10^{-1}	9.9×10^{-2}	4.9×10^{-2}
20	80	2.4×10^{-2}	5.6×10^{-3}	3.0×10^{-4}

the randomization process may be significant during the time of the laser flash, it need not have a noticeable effect on the efficiency of incoherent absorption because [according to the harmonic oscillator model, (3-10)] the absorption coefficient is independent of the molecular distribution among the levels.

However, randomization of the excitation energy *does* change the fraction of highly excited molecules, and hence the capacity of the molecular ensemble to react. When absorption from the laser beam is incoherent, that is, at moderate-to-high pressures, this change is probably a decrease and may be considerable. To illustrate the effect, we consider a specific example, the change in the fraction of molecules with $E \geqslant E_{act}$ when a given amount of vibrational energy, concentrated initially in one vibrational mode, becomes distributed over s vibrational modes. We assume that the oscillator frequencies of the modes are equal, and that both the initial and final energy distributions are equilibrium Boltzmann distributions. Given these assumptions, the statistical calculation is straightforward and has been discussed by Kassel [108]. Assuming oscillator frequencies of 1000 cm^{-1} and representative values of 10 or 20 kcal/mole for E_{abs} and 40 or 80 kcal/mole for E_{act}, results of such calculations are listed in Table 4-1. It can be seen that the fraction of reactive molecules decreases in all cases. The decrease becomes more pronounced as E_{act}/E_{abs} increases, or as the number of oscillators over which E_{abs} becomes distributed increases, and in the examples shown it is as much as four orders of magnitude.

Some Definitions

For an ensemble of molecules in the electronic ground state, the internal energy may be divided into additive contributions due to translational, rotational, and vibrational modes of motion: $E = E_T + E_R + E_V$. Accordingly, in the analysis of energy-flow patterns, it is convenient to discuss the flow of energy within and between the three kinds of modes. The notation used in such discussions is exemplified by the following:

R-R	exchange of rotational energy
R-T	exchange between rotation and translation
V-T	exchange between vibration and translation
T/R	R-T, R-R, and T-T are so facile that the rotational and translational modes are practically in statistical equilibrium.
V-V, T/R	exchange of vibrational energy accompanied by energy exchange with T/R

Exchange of energy among the V, R, and T degrees of freedom requires collisions. If the molecules were truly hard spheres, the concept of a collision would pose no problem. In fact, real molecules are not hard spheres. As a result, molecular radii depend on the process under consideration. For instance, radii based on gas viscosity differ somewhat from those based on X-ray crystallography, and the quantum mechanical theory holds that the full wave functions of molecules extend to infinity.

The vagueness of the concept of molecular radii introduces a corresponding vagueness into the concept of molecular collision. In discussions of collisional energy exchange, it is convenient to introduce as a frame of reference the *gas-kinetic* collision, the familiar event in which molecules exchange vector momentum and experience T-T exchange. For typical molecules in the gas phase at P torr total pressure, the mean time between successive gas-kinetic collisions is about $80/P$ ns.

R-R and T-R Exchange

Relative to the frame of reference (the gas-kinetic collision with T-T exchange), R-R and T-R energy exchanges occur fast, the mean time between successive energy exchange events being only about one-tenth of that between successive gas-kinetic T-T collisions [5, 37]. Thus it may be assumed in good approximation that statistical equilibrium exists between rotation and translation throughout many processes of energy exchange with vibration. The T/R degrees of freedom are called the *heat bath* and, being in statistical equilibrium, may be characterized by a temperature.

V-T Exchange

Quantitative information about the speed of V-T energy exchange (strictly speaking, V-T/R exchange) was derived originally from ultrasonic measurements [38]. In recent years, the ultrasonic results have been confirmed and amplified by a variety of pulsed laser techniques [39].

The probability per collision of V-T exchange decreases dramatically and exponentially with the amount of energy being transferred [40]. This effect is particularly noticeable when the transfer of small energy quanta is prohibited by quantum restrictions, for instance, when the lowest vibrationally excited state exchanges with the vibrational ground state. Energy gaps between the lowest excited state and the ground state vary from less than 100 cm^{-1} for flexible molecules all the way up to over 1000 cm^{-1}. In the case of CH_3F, the lowest vibrationally excited state lies 1049 cm^{-1} above the ground state, and 8000 gas-kinetic collisions are required to transfer this energy [41]. In the case of CBr_2F_2, where the lowest excited state lies 165 cm^{-1} above the ground state, the corresponding number is 70 collisions [42].

V-T exchange is more efficient for hydrogen compounds than for compounds without hydrogen atoms [40]. When molecular structures permit "sticky" collisions, V-T exchange is particularly efficient. Examples are collisions between hydrogen-bond donor and acceptor molecules, collisions between flexible molecules which can become entangled, collisions of uncharged molecules with ions, and collisions between molecules of differing electronic multiplicity, such as a free radical with its parent hydrocarbon, where London dispersion forces are apt to be large [43].

In many cases, the facts concerning V-T exchange efficiency are too specific to be rationalized by any simple generalization. For instance, He is an excellent agent for V-T energy transfer from the lower laser levels in the CO_2 laser, but in other cases it is among the least efficient of energy acceptors.

V-V, T/R Exchange

In infrared laser chemistry, where one is dealing with molecules in very high vibrational states, direct transfer of vibrational energy to translation is less important than the transfer of vibrational energy between the colliding molecules. This process is usually accompanied by exchange of energy with the T/R heat bath. The energy changes involved in V-V, T/R exchange are shown in Fig. 4-2. We may assume that the vibrational energy of at least one of the colliding molecules lies in the quasi-continuum. Because for a *real* continuum there are no quantum restrictions to the addition or removal of energy, one might expect that the final states in V-V, T/R exchange are consistent with the Boltzmann probability prediction [44, 108]. In fact, this is not the case. There is a marked tendency for the amount of energy ϵ being transferred to the heat bath to be small; processes with $\epsilon > 50\ \mathrm{cm}^{-1}$ are rare [39]. Also, there are detailed preferences for certain final vibrational states over others, for which, at the present time, no reliable generalizations exist.

Usually, V-V, T/R exchange is most efficient between molecules belonging to the same substance [44]. The reason for this is that conditions for *resonant* energy transfer can be met. One of the conditions is that $\epsilon \cong 0$. However, that condition can always be met if one of the molecules is highly excited, owing to the great density of vibrational levels. Another condition is, roughly speaking, mechanical. When the vibrational energy is analyzed in terms of normal modes, each mode represents a specific kind of atomic vibration. When a collision takes place, energy transfer is resonant if the vibrations of the colliding molecules permit energy exchange without hindrance. An example of this condition being met is that energy migrates from a given mode of one molecule to the *same* mode of another molecule.

V-V, T/R exchange efficiencies for molecules belonging to different substrates are highly variable, as later examples will show. A common

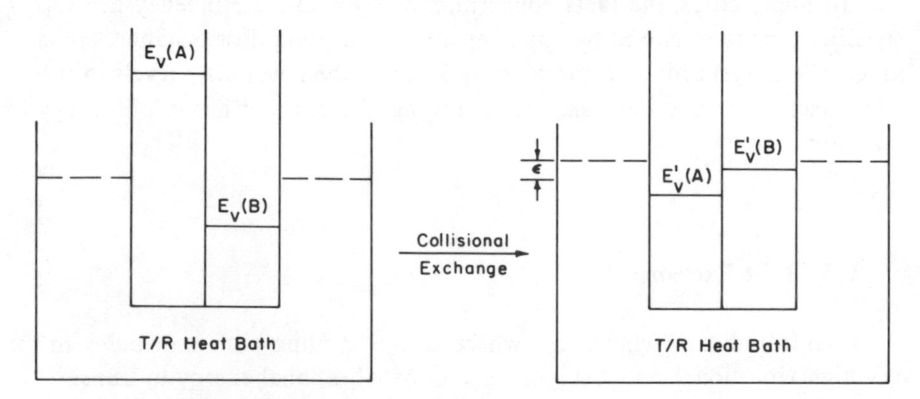

Fig. 4-2 Schematic representation of V-V, T/R exchange.

sequence of relative efficiency is polyatomic > small linear or diatomic > monotomic [44]. Another sequence is hydrogen compounds > nonhydrogen compounds. As in V-T exchange, the concept of "sticky" collisions involving large and/or polar molecules has been invoked in V-V, T/R exchange to explain unusually high efficiencies [45].

SOME EXPERIMENTAL RESULTS

Energy Spoiling by Inert Gases

Addition of an inert gas tends to increase E_{abs} for the reactive substrate (Chapter 3), but this favorable effect is often overwhelmed by a "spoiling effect" which reduces the CPF. Energy spoiling is relevant to the discussion of energy-flow patterns because it proceeds primarily by removal of vibrational energy from the absorbing molecules in collisions. When reaction takes place from a nonequilibrium molecular energy distribution, there is also a subsidiary phenomenon (which will here be neglected), because collisions with the added gas molecules will accelerate the approach to thermal equilibrium and thus change the fraction of activated molecules.

Data for energy spoiling are scarce because the comparisons must be made at constant E_{abs}. Results in Table 4-2 illustrate both the importance of the effect and its specificity. Energy spoiling has been evaluated in two ways. The column CPF/(CPF)$_0$ compares the actual CPF in the presence of the added gas with the value (CPF)$_0$ observed in the absence of the added gas at the same E_{abs}. The column, *quasi*-E_{abs}/E_{abs}, evaluates the fraction of E_{abs} that appears to be spoiled; E_{abs} is the experimental value, and *quasi*-E_{abs} is the amount of absorbed energy which, in the absence of the added gas, would have given the CPF.

The limited available data may be summarized as follows. The *spoiling ratio*, *quasi*-E_{abs}/E_{abs}, due to added gas, depends approximately on the pressure ratio, P_X/P_A, rather than on P_X alone. The spoiling ratio is also approximately independent of E_{abs}. In view of this, one may define a spoil factor f by way of (4-1).

$$\frac{(quasi\text{-}E_{abs})}{E_{abs}} = \frac{1}{\left[1+f(P_X/P_A)\right]} \tag{4-1}$$

Table 4-2 Selected Data for Energy Spoiling by Inert Gases [10, 11]

P_A (torr)	P_X (torr)	E_{abs} (kcal/mole)	$\frac{CPF}{(CPF)_0}$	$\frac{quasi\text{-}E_{abs}}{E_{abs}}$	Spoil factor (f)
	$A = CCl_2F_2$				
12	N_2, 22.1	21.5	0.1	0.665	0.27
12	SiF_4, 4.8	21.5	0.25	0.782	0.70
12	CBr_2F_2, 9.2	21.5	0.6	0.901	0.14
	$A = CHClF_2$				
5.8	He, 29.0	16.5	0.0030	0.388	0.32
5.8	N_2, 11.6	13.1	0.011	0.463	0.58
8.6	SiF_4, 3.9	15.1	0.064	0.626	1.33

In actual cases, f is either constant or only somewhat variable. It provides a measure of effectiveness of spoiling under normalized conditions.

Comparison of results for the two absorbing substrates shows that the hydrogen compound $CHClF_2$ is much more effective at transferring energy to an added gas than is CCl_2F_2, as expected. The polyatomic molecule SiF_4 is more effective at accepting energy than is N_2. The relatively high spoil factor for He is less easy to explain, because energy transfer is limited to the V-T mechanism. Finally, the smallness of the spoil factor of CBr_2F_2 with CCl_2F_2 is surprising in view of the similarity of the two molecules and the near-coincidence of several normal-mode frequencies [46, 47].

Sensitized Reactions

Because in a sensitized reaction, energy must be transferred from the absorbing substance to the desired reactant, the efficiency of V-V,T/R transfer must be high. Thus there is a positive connection between energy spoiling and sensitizing efficiency. For instance, Table 4-2 shows that $CHClF_2$ undergoes V-V,T/R exchange more easily than does CCl_2F_2. Because exchange is reversible, one expects that reactions of $CHClF_2$ will be sensitized more easily than those of CCl_2F_2. This is in agreement with fact, as proved by the data in Table 4-3.

Table 4-3 Comparison of CPF for Unsensitized and SiF_4-Sensitized Reactions

Gas Phase	Absorber	Frequency (cm^{-1})	E_{abs} (J/cm^2)	CPF
	(*A*) $CHClF_2 \rightarrow CF_2 + HCl$[a]			
$CHClF_2$(50 torr)	$CHClF_2$	1088	0.124	23%
$+SiF_4$(23 torr)	SiF_4	1025	0.145	9%
	(*B*) $CCl_2F_2 \rightarrow CClF_2 + Cl$ [11]			
CCl_2F_2(11.5 torr)	CCl_2F_2	921	0.18	0.24%
$+SiF_4$(5.6 torr)	SiF_4	1031	0.40	$<0.01\%$

[a] Data of K. J. Olszyna, private communication, August 1977.

The ideal sensitizer would be inert, absorb strongly at a suitable frequency, and be efficient at V-V, T/R transfer. Unfortunately, hydrogen compounds that satisfy the third requirement are usually reactive. Thus, when a gaseous mixture of hydrogen compounds is irradiated at a frequency where only one of the components absorbs, energy redistribution is quite effective so that all components become activated. For instance, gas-phase samples of ethyl acetate (15 torr) and isopropyl bromide (5 torr) were irradiated at 1050 cm^{-1}, where only the ethyl acetate absorbs [15]. Both substances decompose according to the well-established thermal reactions:

$$CH_3C(=O)OC_2H_5 \longrightarrow CH_3C(=O)OH + H_2C{=}CH_2 \tag{4-2}$$

$$CH_3CHBrCH_3 \rightarrow CH_3CH{=}CH_2 + HBr \tag{4-3}$$

At 4 MW/cm^2, the product ratio agrees with that for thermal reaction, indicating highly efficient energy sharing. However, when the power is increased by an order of magnitude, the relative proportion of acetic acid and ethylene among the products goes up [15, 48]. Qualitatively similar results have been reported for the irradiation of a gaseous mixture of allylmethyl ether and isopropyl bromide, at a frequency at which only allylmethyl ether absorbs [48].

SOME THEORETICAL CONCEPTS

We now analyze two theoretical ideas that are important with respect to energy-flow patterns: (1) the marked reversibility of energy flow, even far from thermal equilibrium, when energy is being transferred in small increments; and (2) the accumulation of vibrational energy in specific modes under some conditions.

Reversibility of V-V, T/R Exchange

Any macroscopic process on the microscopic level is the difference between a forward rate and a reverse rate. In *chemical* rate processes far from equilibrium, the reverse rate may often be neglected. However, in *energy-flow* processes, the reverse rate may not be neglected under any conditions if the average amount of energy per microscopic event is small. This situation often exists in V-V, T/R exchange and has a profound effect on the net speed of the macroscopic process. The following calculation based on a simplified model illustrates this point.

Let ϵ denote the amount of energy transferred to or from the heat bath in a given collision, and let $\tilde{\epsilon}$ denote the root mean square average value of ϵ. Actual values of ϵ will vary according to a specific distribution function. However, for simplicity we shall assume that a fixed amount, $\tilde{\epsilon}$, is exchanged in each collision. This leads to a two-valued distribution function: The probability is p_+ that $\tilde{\epsilon}$ is added to the heat bath, and $p_- = 1 - p_+$ that $\tilde{\epsilon}$ is removed from the heat bath. Let η denote the resultant net energy flowing *into* the heat bath per collision. In terms of this model, η is then given by (4-4).

$$\eta = \tilde{\epsilon}\,(p_+ - p_-) \tag{4-4}$$

To obtain an expression for $p_+ - p_-$, we focus on the vibrational energies of the colliding molecules. As indicated in Fig. 4-3, the process of adding to the vibrational energy from the heat bath requires an activation energy $\tilde{\epsilon}$, while that of removal does not. Letting T denote the temperature of the heat bath, we therefore [108] write

$$\frac{p_-}{p_+} = \frac{A_-}{A_+}\exp\left(\frac{-\tilde{\epsilon}}{kT}\right) \tag{4-5}$$

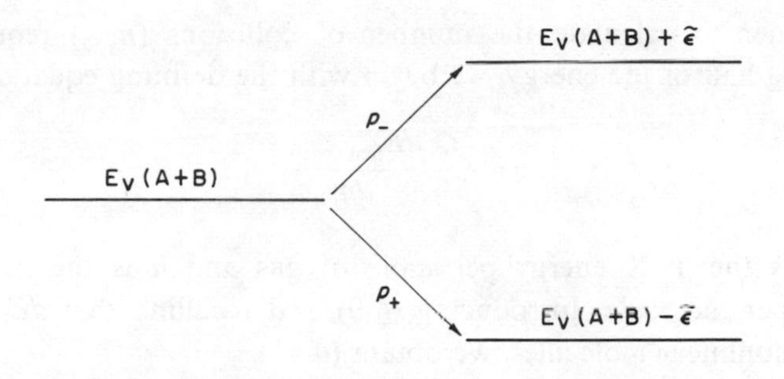

Fig. 4-3 Energy change in model for V-V, T/R exchange. Energy change of heat bath (not shown) is complementary.

When thermal equilibrium exists, $T = T_e$ and $p_-/p_+ = 1$. Thus $A_-/A_+ = \exp(\tilde{\epsilon}/kT_e)$, and we obtain (4-6) and (4-7).

$$\frac{p_-}{p_+} = \exp\left[-\frac{\tilde{\epsilon}}{k}\left(\frac{1}{T} - \frac{1}{T_e}\right)\right] \tag{4-6}$$

$$p_+ - p_- = \frac{1 - \exp[-(\tilde{\epsilon}/k)(1/T - 1/T_e)]}{1 + \exp[-(\tilde{\epsilon}/k)(1/T - 1/T_e)]} \tag{4-7}$$

In V-V, T/R energy exchange, it is usual that $\tilde{\epsilon} \ll kT$. Thus, after the expansion of the exponential term in a power series, (4-7) reduces to (4-8):

$$p_+ - p_- = \frac{\tilde{\epsilon}}{2k}\left(\frac{1}{T} - \frac{1}{T_e}\right); \qquad \tilde{\epsilon} \ll kT \tag{4-8}$$

Finally, the desired energy quantity η is obtained by introducing (4-8) into (4-4).

$$\eta = \left(\frac{\tilde{\epsilon}^2}{2k}\right)\left(\frac{1}{T} - \frac{1}{T_e}\right) \text{ per collision}; \qquad \tilde{\epsilon} \ll kT \tag{4-9}$$

According to (4-9), the net energy flow into the heat bath per collision is remarkably small when $\tilde{\epsilon} \ll kT$. For instance, using typical values of $\tilde{\epsilon} = 10\ \text{cm}^{-1}$, $T = 300°\text{K}$, and $T_e = 1100°\text{K}$, η is only $0.17\ \text{cm}^{-1}$ per collision (0.50 cal/mole of collisions).

In order to calculate the number of collisions ($n_{1/2}$) required for transferring half of the energy, we begin with the defining equation (4-10).

$$\eta = \frac{1}{N_0} \frac{dE_{\text{heat bath}}}{dn} \tag{4-10}$$

$E_{\text{heat bath}}$ is the T/R energy per mole of gas and n is the number of collisions per molecule. Introducing (4-9) and recalling that $dE_{\text{heat bath}} = 3RdT$ for nonlinear molecules, we obtain (4-11).

$$\frac{N_0^2 \tilde{\epsilon}^2}{2R}\left(T^{-1} - T_e^{-1}\right) = 3R\frac{dT}{dn} \tag{4-11}$$

Thus $n_{1/2}$ is given by (4-12), where T_i is the initial heat-bath temperature and $T_{1/2} = (T_e - T_i)/2$.

$$n_{1/2} = \frac{6R^2}{N_0^2 \tilde{\epsilon}^2} \int_{T_i}^{T_{1/2}} \frac{dT}{\left(T^{-1} - T_e^{-1}\right)} \tag{4-12}$$

Representative results are shown in Table 4-4. On the basis of available data for molecules with small excitation energies [38–40], it is unlikely that in the absence of chemical reaction,* values for $\tilde{\epsilon}$ above 20 cm^{-1} need be considered. For highly excited molecules, values for $\tilde{\epsilon}$ may be even smaller because quantum restrictions on energy transfer owing to the discrete spacing of vibrational levels become less stringent.

Table 4-4 shows that the time for V-V,T/R energy half-transfer increases with T_e, and hence with the initial vibrational excitation energy (E_{abs}). Relevant T_e values in megawatt infrared laser chemistry range upwards of 900°K. It is therefore probable that several thousand collisions are needed merely to go half-way to thermal equilibrium. (At a pressure of 50 torr, the time required for 3000 collisions is ~5 μs.) The number, $n_{1/2}$, is so great owing to the high reversibility of the V-V,T/R process. For instance, line 1 of Table 4-4 shows that when $\tilde{\epsilon} = 10$ cm^{-1}, 3600 collisions are needed to effect a net energy transfer of only $21\tilde{\epsilon}$, because $p_+ - p_-$ is only 0.006. The apparent weakness of the coupling between V and T/R modes is especially important in infrared laser chemistry with *continuous*

*In exothermic reactions, a substantial fraction of the reaction exothermicity may appear as T/R energy of the product molecules.

Table 4-4 Number of Collisions Required for V-V, T/R EnergyHalf-Transfer According to (4-12)

T_i	T_e	$T_{1/2}{}^a$	Net Energy transferred (cm^{-1})^b	$n_{1/2}$
		$\tilde{\epsilon}=10\ cm^{-1}$		
300	500	400	210	3.6×10^3
300	700	500	420	5.8×10^3
300	900	600	630	8.4×10^3
300	1100	700	840	11.6×10^3
		$\tilde{\epsilon}=20\ cm^{-1}$		
300	500	400	210	0.89×10^3
300	700	500	420	1.45×10^3
300	900	600	630	2.11×10^3
300	1100	700	840	2.89×10^3

$^aT_{1/2}=(T_e-T_i)/2$.
$^b3R(T_{1/2}-T_i)$ calories per mole.

lasers, where, owing to the relatively slow rate of optical excitation, excess energy will accumulate in the vibrational manifolds only if V-V, T/R exchange is relatively slow.

Accumulation of Vibrational Energy in Specific Modes

When the vibrations of molecules are represented in the harmonic oscillator approximation, an explicit and fairly simple picture of the energy distribution among the vibrational modes can be derived. For definiteness, consider a mixture of two diatomic species A and B such as N_2 and CO. Energy is being introduced in the vibrational ladder of A and then flows as a result of collisions both into the vibrational ladder of B and into the T/R heat bath common to both species. For harmonic oscillators certain assumptions can be made concerning the average speeds of such energy flows.

1. *Energy redistribution up and down a given vibrational ladder is relatively very fast and establishes a Boltzmann distribution with a characteristic temperature (T_A, T_B) within each ladder* [49]. Collisional transfer within

each ladder is resonant and proceeds with high probability to the adjacent level in the ladder. For a molecule in the jth level, the probability per second of accepting a quantum in a collision is (4-13); that of losing a quantum is given by (4-14) [38].

$$p_{j\to j+1} = Z_{01}P(j+1)\bar{v} \text{ per second} \tag{4-13}$$

$$p_{j\to j-1} = Z_{01}Pj(\bar{v}+1) \text{ per second} \tag{4-14}$$

In these equations, $\bar{v}$ is the average excitation number of the molecules occupying the given vibrational ladder, and Z_{01} is a characteristic constant for the vibrational ladder and is typically of the order of 10^9 atm^{-1}/sec [50, 51], only slightly less than the gas-kinetic T-T collision frequency. The equations show that the probabilities are proportional to $\bar{v}$ or $(\bar{v}+1)$. Thus, as the vibrational mode gains excitation energy, collision-induced energy exchange up and down the vibrational ladder accelerates.

2. *Exchange of energy between the two vibrational ladders is slower than that within the ladders, but faster than energy exchange with the T/R heat bath* [*39*]. Energy exchange probabilities are rarely greater than 0.1 per gas-kinetic T-T collision, and may be considerably smaller.

Let us suppose that an energy pulse is introduced into ladder A. The following sequence of events ensues. The A mode quickly comes to a Boltzmann equilibrium with temperature T_A. Energy flows more slowly from mode A into mode B. During this process, the temperature T_B rises while T_A drops, until a stationary state is reached. Energy flows from both A and B into the T/R heat bath, but relatively very slowly. Thus, after an initial transient period, energy flows into the T/R heat bath from vibrational modes of A and B that are in a stationary state with each other.

It can be shown that, in this stationary state, the temperatures T_A and T_B are related to the temperature T of the heat bath by (4-15) [49, 52, 53], where ν_A and ν_B are the harmonic oscillator frequencies. One solution to (4-15)

$$\frac{\nu_A}{T_A} - \frac{\nu_B}{T_B} = \frac{\nu_A - \nu_B}{T} \tag{4-15}$$

is that $T_A = T_B = T$, corresponding to thermal equilibrium. Of greater interest are the temperature conditions in the stationary states preceding thermal equilibrium. The following will show that according to (4-15), energy tends to accumulate in the lower frequency mode.

For instance, consider a mixture of N_2 and CO. For N_2, $\nu_A = 2360$ cm^{-1} and for CO, $\nu_B = 2168$ cm^{-1}. Let $T = 300°K$ and $T_A = 1000°K$. Then

$$\frac{2360}{1000} - \frac{2168}{T_B} = \frac{192}{300}$$

Thus $T_B = 1260°K$.

The accumulation of energy in the lower-frequency mode becomes more striking as the deviation from thermal equilibrium increases. Letting $T = 300°K$ and $T_A = 2000°K$, we find that $T_B = 4015°K$.

To derive stationary-state conditions for polyatomic molecules using the same harmonic oscillator approximation, a set of simultaneous equations must be written such that (4-15) is satisfied for any pair of modes, and that the energy of all the modes adds up to the known total [54]. In the case of $^{13}CH_3F$, it has been shown that this model gives a fairly good representation of the facts [55]. Experimental data for total E_{abs} from a

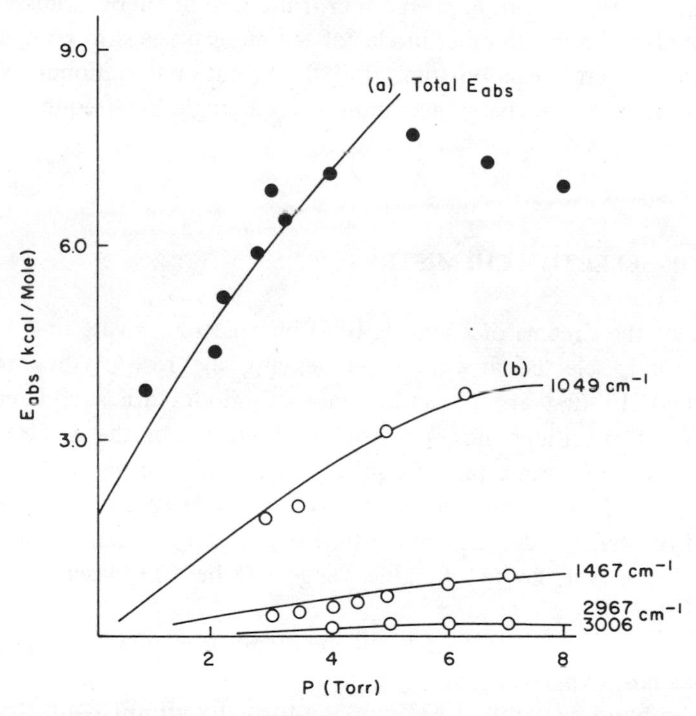

Fig. 4-4 E_{abs}/mole vs pressure of $^{13}CH_3F$ after irradiation at 1049 cm^{-1}. (*a*) Total E_{abs}; (*b*) Energy found in given vibrational mode when stationary state is reached. Based on data in [55].

pulsed CO_2 laser, and of energy stored in specific vibrational modes in the subsequent stationary state (measured by infrared fluorescence) are shown in Fig. 4-4. The solid lines show theoretical prediction along the lines of (4-15) [55]. As expected, most of the vibrational energy is stored in the lowest frequency mode at 1049 cm^{-1}.

It is sometimes found [9–11] that in pulsed megawatt laser chemistry, CPF for pure gases varies with E_{abs} according to a simple exponential relationship (4-16).

$$CPF = A \exp\left(\frac{-E_{act}}{E_{abs}}\right) \tag{4-16}$$

E_{act} is the activation energy for the primary reaction step; the pre-exponential factor A is of the order of unity. A plausible interpretation of (4-16) is that reaction takes place from a Boltzmann distribution in which the absorbed energy is concentrated in one vibrational mode [9]. In terms of present theory, this can happen either if the rate of energy transfer from the mode of excitation to other modes of the molecule is slow compared to that of the primary reaction step, or if the vibrational stationary state is such that much of the energy accumulates in a single low-frequency mode [56].

MODE-SELECTIVE CHEMISTRY

One of the dreams of chemists is to be able to activate and break a specific bond, selected at will, while leaving the rest of the molecule undisturbed. In first approximation, normal modes are associated with vibrations of specific bonds, or groups of atoms, in the molecule. By choosing the appropriate laser frequency, such specific motions could be excited. Thus, if (or when) a continuously tunable laser becomes available, this kind of research can be done efficiently. However, experiments done to date within the limited tunable range of the CO_2 laser have been disappointing.

An example is the laser-induced decomposition of CCl_2F_2 [11]. This substance has two strong absorption bands in the CO_2 laser's tunable range. The band centered at 923 cm^{-1} is formally an antisymmetric CCl_2 stretch; that centered at 1098 cm^{-1} is formally a symmetric CF_2 stretch. The word "formally" is used in this connection because the actual motions are more complicated.

At the same time, there are two parallel reaction channels, (4-17) and (4-18), whose activation energies differ by less than 10 kcal/mole.

$$CCl_2F_2 \rightarrow \cdot CClF_2 + Cl\cdot \tag{4-17}$$

$$CCl_2F_2 \rightarrow :CF_2 + Cl_2 \tag{4-18}$$

Possible conditions for mode-selective reactions therefore exist.

In fact, the reaction products are independent of the mode being excited, and both CPF and product distribution depend only on E_{abs}. Experimental results for CPF vs E_{abs}^{-1} are shown in Fig. 4-5. Within the experimental error, all data are reproduced by a straight line whose

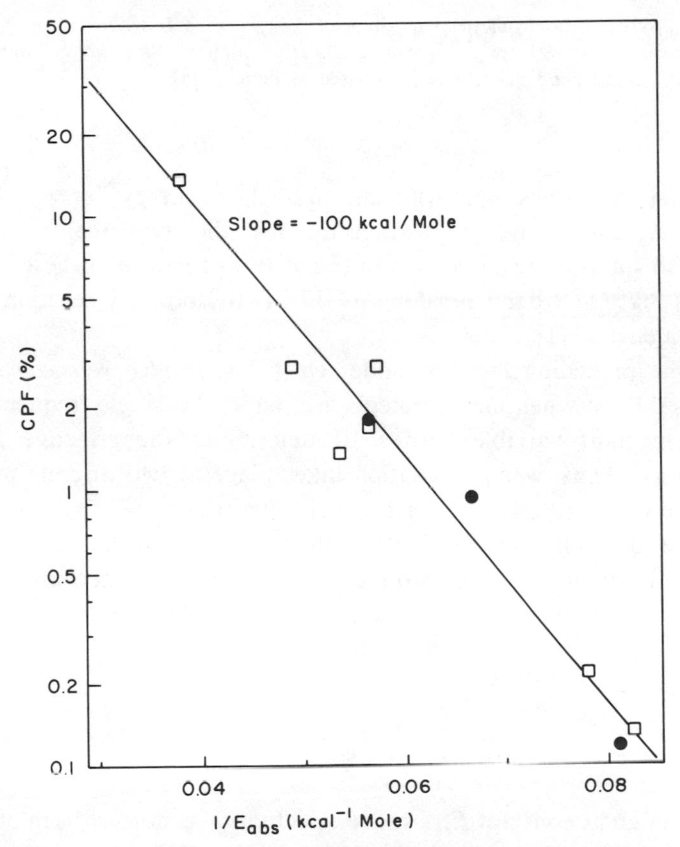

Fig. 4-5 CPF vs $1/E_{abs}$ for laser-induced decomposition of CCl_2F_2 at 921 cm^{-1} (closed circles) and 1088 cm^{-1} (open squares). Straight line of semilogarithmic plot having slope of −100 kcal/mole represents points at both frequencies. Based on data in [11].

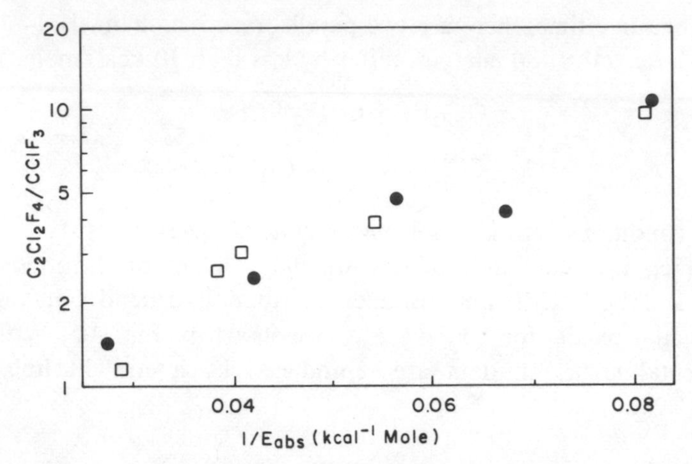

Fig. 4-6 Product ratio from decomposition of CCl_2F_2 as a function of $1/E_{abs}$. Closed circles, irradiation at 921 cm^{-1}; open squares, irradiation at 1088 cm^{-1}. Note that the product ratio is independent of frequency. Based on data in [11].

negative slope agrees well with the activation energy for pyrolysis of CCl_2F_2. A corresponding relationship for the two major products, $CClF_2CClF_2$ and $CClF_3$, is shown in Fig. 4-6. The nature of these products indicates that C-Cl bond breaking [(4-17)] is the primary reaction step at both frequencies [11].

In the preceding test for mode selectivity, *results were compared at equal* E_{abs}. Even when measurements are made at a single frequency, E_{abs} is an important variable because it determines the effective reaction temperature. Thus, when irradiation takes place at two discrete modes, a difference of as little as 20% in the respective values of E_{abs} can cause a difference of well over 100°C in effective reaction temperature. The resulting difference in rate, product ratios, and even mechanistic paths may vitiate any interpretation in terms of mode selectivity.

Molecular Trajectories on a Potential Surface

Although at constant E_{abs}, mode selectivity has not yet been observed, there are reasons why research should be continued. A necessary condition for mode-selective chemistry is that the molecules, after activation, "re-

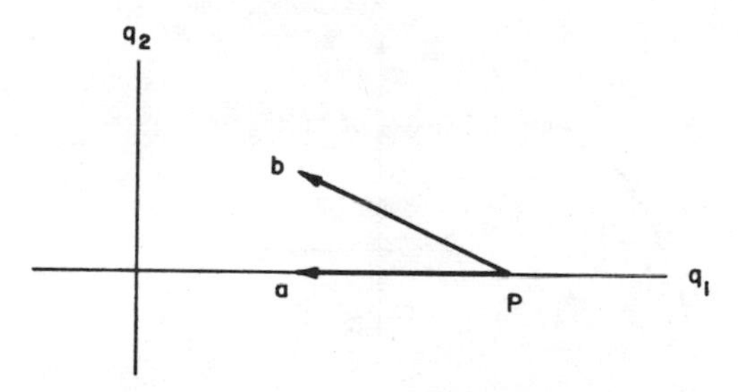

Fig. 4-7 Vector diagram of forces to illustrate coupling between vibrational modes. (*a*) No coupling. (*b*) Coupling.

member" the mode of excitation. This requires that coupling between the absorbing mode and other modes of motion be negligible or at most weak [57, 58] The requirement is most nearly met by stretching vibrations of O-H, C-H, or other bonds to a hydrogen atom. Such vibrations are coupled only weakly to the motions of the rest of the molecule because the light hydrogen atom, which is on the periphery of the molecule, does practically all of the vibrating.

To visualize coupling of vibrational modes, consider a two-mode oscillator with normal coordinates q_1 and q_2. According to Fig. 4-7, assume that the system is momentarily at rest at point P, where the displacement from equilibrium is purely along q_1. If the potential energy consists of separate terms due to q_1 and q_2, the restoring force is a vector towards the origin from P. If it also contains cross-terms involving both q_1 and q_2, the actual force will have a component directed so as to excite q_2. Potential energy functions of polyatomic molecules at higher energy values always contain coupling terms.

The vibration of a molecule as a function of time may be represented by the trajectory of a point on the potential surface. This surface is multidimensional, because the potential energy is a function of the *n* normal-coordinate displacements. We humans find it difficult to visualize more than three dimensions, and our intuition for comprehending the intricacies of multidimensional travel is therefore limited.*

For instance, even such a simple molecule as CCl_2F_2 requires 10-dimensional space $[V(q_1, q_2, \ldots, q_9)]$. That kind of space is unimaginably complicated, and simple analogies to three- or even two-dimensional space

*"How awkward is the human mind in divining the nature of things, when forsaken by the analogy of what we see and touch directly."—Ludwig Boltzmann

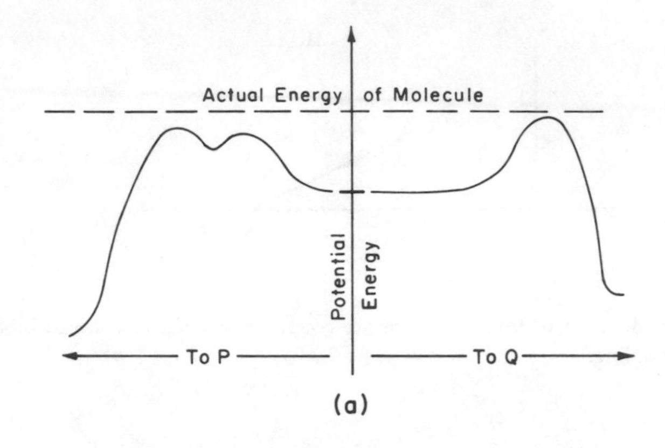

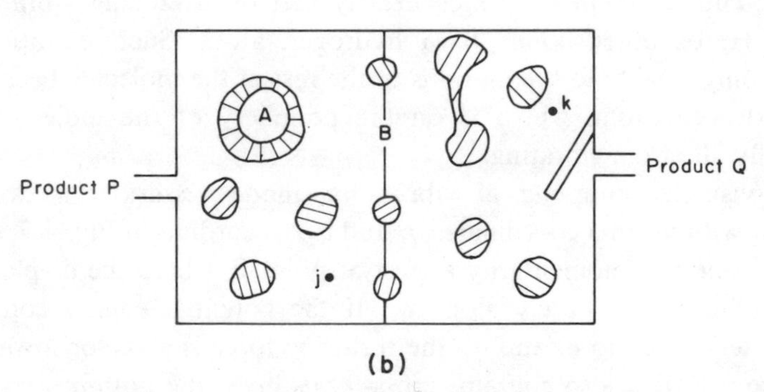

Fig. 4-8(*a*) Potential energy diagram for two parallel processes. (*b*) Two-dimensional representation of potential space. Region *A* is a potential trap and line *B* is a barrier; the two possible reaction paths are shown as narrow channels leading to product *P* or product *Q*. Starting points for the trajectories of the activated molecules are represented by *j* and *k*.

may not hold up. Bearing these caveats in mind, let us examine diagrams that attempt to show, in two dimensions, conditions under which mode-selective chemistry might be possible.

Figure 4-8*a* is a familiar potential energy-reaction coordinate diagram, drawn to depict two parallel reactions of the substrate. Suppose that the actual energy of the molecule is fixed at the level shown, which is higher than the activation energy for either process. Figure 4-8*b* then tries to depict the possibilities for travel of such a molecule on the potential surface. The solid areas represent potential mountains that are inaccessible at the given amount of energy. The two reaction paths are shown as narrow exit channels. The region *A* is a potential trap; once in it, the molecule must experience an activating collision in order to get out. Line *B* represents a barrier between the two reaction channels.

Mode selectivity requires that barriers such as *B* exist. A molecule arriving at point *j* will then go on to product *P*, while one arriving at point *k* will go on to product *Q*. However, the "barrier" need not be an actual energy barrier. It may be a probability barrier, in the sense that a molecule arriving at *j* is so much "closer" to the exit channel *P*, and the anharmonic coupling forces driving it towards channel *Q* are so relatively weak, that the likelihood of its reaching channel *Q* is quite negligible.

The nature of molecular trajectories is poorly understood. One theory (RRKM) holds that potential traps (such as region *A*) or other travel-impeding features of potential surfaces are unimportant [58, 59]. On the other hand, theoretical analysis of plausible potential surfaces indicates that potential traps do exist, especially at intermediate values of the energy [60]. Other travel-impeding possibilities have been discussed [61]. There seem to be enough gaps in theoretical understanding that experimental searches for mode-selective chemistry should be pressed vigorously, especially when such searches will elucidate the characteristics of potential surfaces.

Chapter 5

Practical Chemistry

In order for infrared laser chemistry to become practically useful, it must be economical, controllable, and offer specific advantages over more conventional chemical methods. In this chapter, we give a qualitative discussion of these matters and suggest the broad scope of infrared laser chemistry by reviewing some actual reactions.

ECONOMICS

The cost of infrared photons produced by an efficient CO_2 laser on a large industrial scale has been estimated to be between 2 and 5¢/mole by economists working independently at Exxon Research & Engineering Company and at Allied Chemical Corporation. This estimate includes cost of capital, depreciation, maintenance and supplies; it is believed to be conservative [62].

In order to estimate the cost per mole of laser-induced reaction, we must make realistic estimates of the efficiency with which such photons are used. Let us assume that one-half of the photons are actually absorbed by the reactant and that the photochemical efficiency, defined as the fraction of absorbed photons which is utilized as activation energy, is also one-half. Let us assume an activation energy of 80 kcal/mole, which is equivalent to approximately 28 photons. To convert 1 mole of reactant to product, the laser output actually required then is 4×28, or 112 moles of photons. At between 2 and 5¢/mole, this calls for an outlay of between $2.00 and $6.00. For the kind of reactions that have been studied, molecular weights have been on the order of 100. The production cost is therefore between $20,000 and $60,000 per metric ton, exclusive of the cost of starting materials.

By comparison, the current production cost of penicillin-G, the product of an expensive fermentation process, is about $20,000 per ton. This comparison shows that laser chemistry at this time is too expensive for industrial use, except for special applications. Some applications that are actually being considered are removal of part-per-million amounts of environmentally objectionable materials from industrial wastes. Other industrial applications that have been suggested [62, 63] include treatment of surfaces, generation of radicals of a wide variety, regeneration of poisoned catalysts, and production of catalytically active colloidal particles by homogeneous pyrolysis.

Improvement of Photochemical Efficiency

Although there are examples where the laser energy is used as random thermal energy [14], it is clear that in many cases it is used much more selectively and efficiently. By appropriate tuning of the laser, energy is introduced into the vibrational manifold of a specific reactant and, in some cases at least, stays there long enough to be largely or entirely effective as activation energy.

For instance, in the laser-induced decomposition of CCl_3F, an experiment has been reported in which $E_{abs} = 13.3$ kcal/mole and CPF = 2.8% [9]. The primary process is given in (5-1) and has

$$CCl_3F \rightarrow :CClF + Cl_2 \qquad (5\text{-}1)$$

an estimated activation energy of 81 kcal/mole. The photochemical efficiency is therefore $(2.8 \times 81)/13.3 = 17\%$. If E_{abs} had been converted entirely to random thermal energy, the efficiency would have been smaller by several orders of magnitude. Efficiencies greater than 50% are not uncommon for megawatt laser-induced reactions [10].

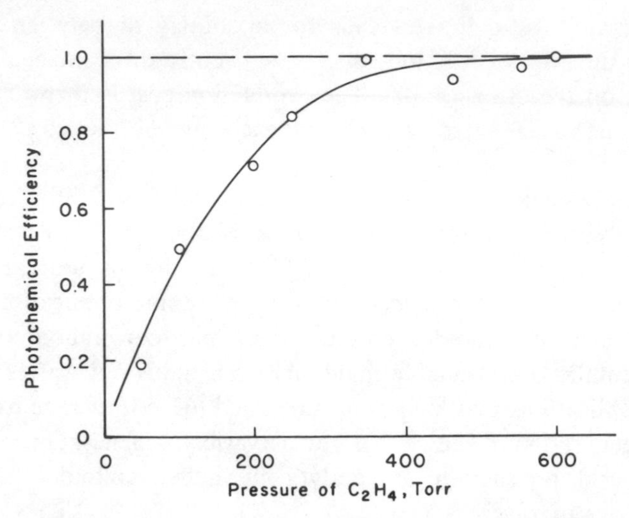

Fig. 5-1 Photochemical efficiency vs pressure in the decomposition of C_2H_4. A continuous laser operating at 943 cm^{-1} with a power of 660 W/cm^2 was used. The quantum efficiency approaches unity as the pressure is increased. Based on data in [65].

Photochemical efficiency has been measured also for reactions induced by *continuous* lasers [64, 65]. For example, the laser-induced pyrolysis of ethylene produces essentially the same products that result from thermolysis—H_2, CH_4, C_2H_6, C_2H_2, propylene, butene, and butadiene—as well as some carbon tars [64]. The overall activation energy for the disappearance of ethylene is about 45 kcal/mole. For the laser-induced process at 943 cm^{-1} and 660 W/cm^2, the photochemical efficiency is shown in Fig. 5-1 as a function of C_2H_4 pressure. At low pressures it is well below unity and strongly pressure-dependent, but at pressures above 300 torr the value approaches unity.

The defining equation for photochemical efficiency suggests a limiting value of unity, corresponding to full use of E_{abs} in overcoming the activation barrier. However, this suggestion is misleading, because the energy that is liberated when the activated complex goes on to products may be returned to the reactant molecules in collisions and used as additional energy of activation. An alternate measure of laser-chemical efficiency is *quantum requirement*, defined as the number of photons absorbed per product molecule produced. Experimental results [65] obtained for the quantum requirement in the laser-induced decomposition

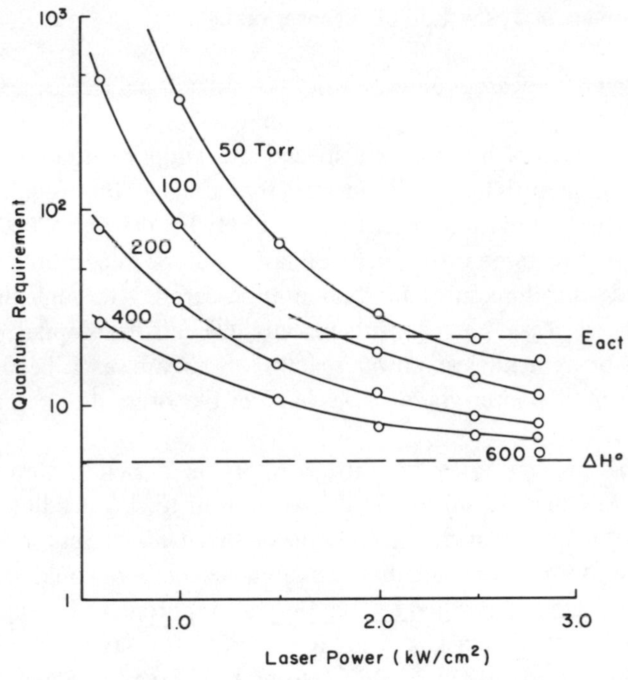

Fig. 5-2 Quantum requirement vs laser power for CW laser-induced dehydrohalogenation of CH_3CClF_2 at 953 cm^{-1}. Note that quantum requirement decreases with increasing pressure, passing below E_{act} for all pressures and approaching the endothermicity $\Delta H°$ at higher pressures. Based on data in [65].

[given in (5-2)] of CH_3CF_2Cl are shown in Fig. 5-2. Continuous laser intensities

$$CH_3CF_2Cl \rightarrow CH_2{=}CF_2 + HCl \tag{5-2}$$

at 953 cm^{-1} ranged up to 2.8 kW/cm^2. The figure shows that the quantum requirement is even less than the activation energy under many conditions, and that it tends toward a limiting value equal to the endothermicity of the reaction $\Delta H°$.

Data such as those in Figs. 5-1 and 5-2 show that the cost of laser-induced reaction can be reduced substantially below the estimate of $20,000–$50,000 per ton, not only by reduction in the cost per mole of infrared photons through laser development, but also through improvements in photochemical efficiency.

Economics of Laser Isotope Separation

Although research interest in laser isotope separation extends throughout the periodic table, the economics of electrical power generation by nuclear reactors has focused special attention on uranium and deuterium enrichment [66]. It is estimated that nuclear fuel requirements for this planet will reach 6 million tons of U_3O_8 by the year 2000. Thus the influence of successes in the field of laser isotope separation will be felt well outside the domain of fundamental research. Although the uranium enrichment process now consumes only 6% of the capital investment needed to bring a nuclear power reactor on stream, as it becomes necessary to turn to poorer-grade or less-accessible ores, this percentage will surely rise.

At the present time, gas-diffusion plants enrich uranium as UF_6, initially 0.75% in ^{235}U, up to 3%; the waste sent to slag is still 0.25% in the desired isotope. One of the attractions of the use of lasers in enrichment technology is that the potential for high selectivity would increase the fraction extracted; this could reduce the ore requirement by approximately 40%. Another attraction is the great potential for saving in energy cost. Gas diffusion costs about 5 MeV/atom(!), whereas the cost of infrared laser isotope separation is estimated to be about 10 eV/atom.

Unfortunately, there are still formidable technical difficulties to be overcome. The frequency difference between $^{235}UF_6$ and $^{238}UF_6$ is less than 1 cm^{-1} in the 624 cm^{-1} band where the gas absorbs strongly. It is true that the monochromaticity of infrared laser-emission peaks is much better than this, being within 0.05 cm^{-1} in many cases. However, there are two impediments to the immediate application of infrared lasers to isotope separation of heavy metals: (1) The half-widths of infrared absorption bands are considerably greater than their isotopic frequency shifts, and (2) in the presence of very intense radiation fields, they become even broader. However, two-frequency laser techniques are being developed that show promise of solving these problems. (See the final section of this chapter.)

As a matter of fact, isotope separation for nuclear power may be applied to the production of heavy water before it is applied to that of ^{235}U. In contrast to the light-water reactor which uses ^{235}U-enriched fuel, the heavy-water reactor uses natural uranium. For the latter, the cost of D_2O amounts to 25% of the total capital outlay. The potential economic benefits to be derived from inexpensive laser-D_2O separation are therefore considerable. The technical difficulties are of a different type, since the frequency difference between H_2O and D_2O in the infrared is so great as to cause no problem of isotope-specific excitation. What needs to be developed is an infrared laser that emits at the proper frequency with high enough power, and which also can be scaled up to industrial size. At this time, the outlay for the distillation-produced D_2O for a typical heavy-water reactor is $100 million; a substantial reduction in this cost would make the heavy-water reactor truly economical.

CONTROLLABILITY

The previously described division of infrared laser chemistry into gigawatt, megawatt, and low-power CW regions recognizes the fact that chemical behavior depends very much on E_{abs}, which in turn depends on laser parameters such as power, dose, and power profile, which vary in these regions. The potential of infrared laser chemistry is revealed perhaps most clearly if we think of the laser beam as being analogous to a high-energy reagent in chemical synthesis. However, in contrast to the chemical reagent whose molecular structure and therefore strength can be changed only in discrete steps, the infrared laser dose can be changed continuously and its chemical effectiveness fine-tuned by changes in power profile and frequency.

Analogy between E_{abs} and RT

To allow our chemical intuition to go to work, it is useful to draw an analogy between E_{abs} in laser chemistry and RT in high-temperature chemistry. Empirically, ln(CPF) varies as $1/E_{abs}$ (4-16), just as *ln(k)* varies as $1/RT$. Thus, high values of E_{abs} are analogous to high-temperature conditions and favor high-energy reaction paths. The analogy is not exact, because laser-induced reactions take place from a range of molecular energy distributions, depending on conditions whose thermodynamic temperature is not defined.

Continuing with this parallel, a marked change in E_{abs} (for instance, a doubling) is equivalent in its effect to a *roughly* proportional change in RT.

For instance, when E_{abs} changes from 15 kcal/mole to 30 kcal/mole, the chemical effect may be analogous to that of raising the reaction temperature from 1000 °K to 1500–2000 °K. The chemical consequences thus may be dramatic: The dominant reaction path may change to one of considerably higher activation energy (especially since higher-energy processes tend to have higher activation entropies); primary reaction products that are stable at the smaller E_{abs} may decompose and be entirely absent from the isolated products at the greater E_{abs}.

Increase of E_{abs} with Dose

In many of the examples to be described, the investigators do not report E_{abs} but *do* measure the dose. Fortunately, when pulsed lasers are used, it is safe to assume that E_{abs} increases monotonically with the dose, other things being equal. Genuine saturation is rare, even at gigawatt power levels.

For instance, Fig. 5-3 shows E_{abs} as a function of both dose and average intensity I_D for pulsed irradiation of 1 torr of trans-1, 2-dichloroethylene [67]. Dose and average intensity vary by four orders of magni-

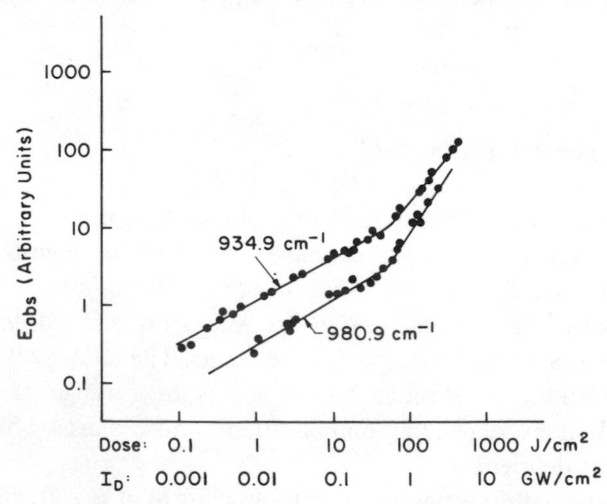

Fig. 5-3 E_{abs} vs dose and I_D for *trans*-1,2-dichloroethylene. Based on data in [67]. Note the change in slope near 0.5 GW/cm^2.

tude. Below 1 GW/cm^2, E_{abs} is roughly proportional to $I_D^{1/2}$; above 1 GW/cm^2, the variation becomes proportional to the first power. One of the irradiation frequencies, 934.9 cm^{-1}, lies in the R branch of the very strong infrared-active C-H bending mode. The other frequency, 980.9 cm^{-1}, does not correspond to any normal mode; it *does* come close to a combination frequency of the infrared-active torsion mode at 227 cm^{-1} with the infrared-inactive bending mode at 763 cm^{-1}. The figure shows clearly that there is no saturation, and that combination frequencies can produce efficient excitation at gigawatt power levels.

Because saturation in absorption from pulsed lasers is rare, we may assume, in the absence of direct evidence to the contrary, that marked increases in dose are attended by marked increases in E_{abs}.

Change in Nature of Products

A convenient way to obtain a high dose from a given laser is by focusing the beam. Although the focal point is not a true geometric point, with good optics the cross-sectional area of the beam at the focal point can be reduced to 3×10^{-3} cm^2 [67]. For a typical laser beam whose area is a few square centimeters, the dose can therefore be increased by as much as three orders of magnitude. Focusing is commonly done with a lens or with a concave mirror (Fig. 5-4).

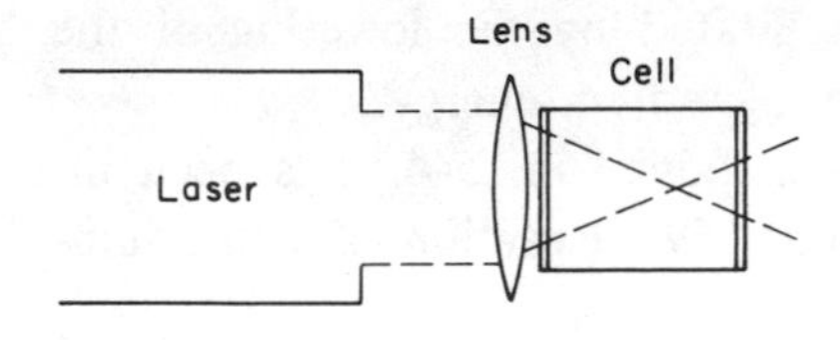

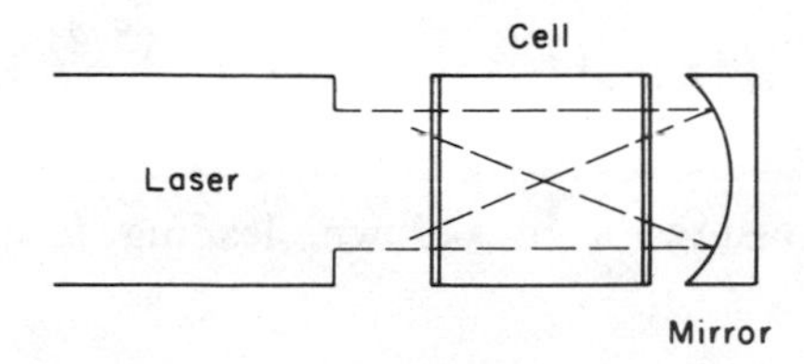

Fig. 5-4 Schematic diagram for experimental focusing of the laser beam.

With focused beams, E_{abs} is substantially greater than with unfocused beams. The effect of this on the nature of products is illustrated by the laser-induced reactions of hexafluorocyclobutene [68]. With unfocused beams, clean conversion to the less stable isomer, hexafluorobutadiene, is observed. With focused beams, the reaction product consists not only of hexafluorobutadiene but also of decomposition products and of low-molecular-weight polymers, as shown in (5-3).

$$\text{hexafluorocyclobutene} \xrightarrow{\text{unfocused}} CF_2{=}CF{-}CF{=}CF_2 \qquad (5\text{-}3)$$

$$\xrightarrow{\text{focused}} C_4F_6\text{, decomposition products (e.g., } C_2F_4\text{), and low-molecular weight polymers}$$

In this case, the experiments do not indicate whether decomposition takes place as a parallel reaction or subsequent to formation of the diene.

One reason that reaction products may be more diverse at the gigawatt per square centimeter intensities characteristic of many focused beams is that dielectric breakdown, leading to ionic products, becomes a distinct possibility. In theory, liability to dielectric breakdown increases with the gas pressure because the "effective ionizing field" at fixed infrared intensity varies approximately as the pressure [21]. Moreover, the ionization of infrared-absorbing species is facilitated by the lowering of the ionization potential due to the absorption of radiant energy.

For instance, the Cope rearrangement [given in (5-4)] has been induced by irradiation with focussed beams [69]. Reaction of deuterium-labeled 1,5-hexadiene is

$$a{=}b{-}c{-}d{-}e{=}f \longrightarrow b{=}c{-}b \ldots \quad (\text{a=b, e=f} \rightarrow \text{b=c, d=e}) \qquad (5\text{-}4)$$

clean at 5–16 torr, but at higher pressures a breakdown, leading to acetylene formation, was observed.

E_{abs} as a Controllable Variable

To the majority of chemists whose practical experience is limited to a temperature range centered on room temperature, the idea that E_{abs} (and hence the effective temperature) can be deliberately varied over several orders of magnitude is unfamiliar, and requires a reorientation of our thinking.

In conventional chemistry, the average internal energy of the reactants and products is only a small fraction of the relevant activation energy, and considerations of the activation energy are therefore paramount. By contrast, in infrared laser chemistry, the internal energy may become comparable to the activation energy and thus gain coordinate importance.

This point is stressed in Fig. 5-5. The figure shows zero-point energy levels E_R°, E_{TS}°, and E_P° for the reactant, transition-state complex and product in a chemical reaction. However, it amplifies this conventional approach by showing also the average internal energy, which for the reactant is close to E_{abs}. Because reaction may take place from a non-equilibrium molecular energy distribution, prediction of the average internal energy in the transition state $(\bar{E}_{int})_{TS}$ is uncertain. However, it seems

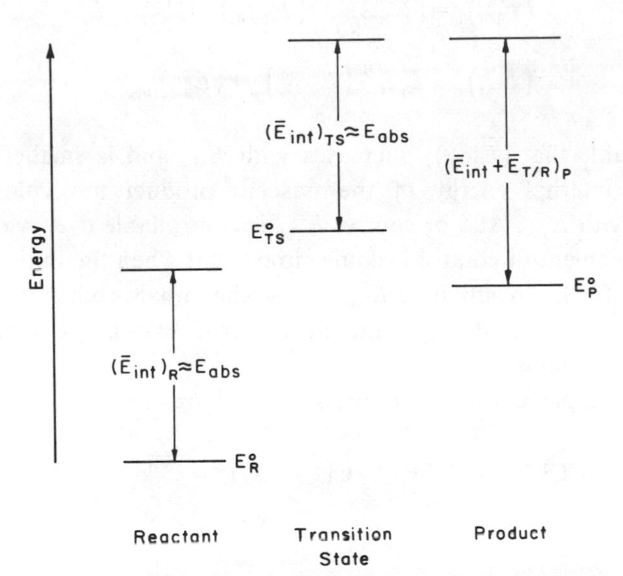

Fig. 5-5 Variation of internal energy in a laser-induced reaction. The variable controlled by the chemist is E_{abs}.

reasonable to assume that $(\bar{E}_{int})_{TS}$ is comparable to E_{abs}, the value for the reactant, and this assumption is made in Fig. 5-5. If the internal energy distribution of the reactant were a Boltzmann distribution according to (3-17), the mean excitation number $\bar{v}_l$ for all molecules with $k \geqslant l$, where l is any positive integer, would be precisely $\bar{v}$, and the assumption would be exact.

$$\bar{v}_l = \frac{\sum_{k=l}^{\infty} kf_k}{\sum_{k=l}^{\infty} f_k - l}$$

$$= \bar{v}, \text{ according to (3-17).} \tag{5-5a}$$

The internal energy of the product is then given by (5-5b), and approximately by (5-5c), where $(\bar{E}_{T/R})_P$ is the mean T/R energy of the nascent product molecules, and $(E^\circ_{act})_{reverse}$ is the zero-point activation energy for the back-reaction.

$$(\bar{E}_{int})_P = (\bar{E}_{int})_{TS} - (\bar{E}_{T/R})_P + E^\circ_{TS} - E^\circ_P \tag{5-5b}$$

$$(\bar{E}_{int})_P \approx E_{abs} - (\bar{E}_{T/R})_P + (E^\circ_{act})_{reverse} \tag{5-5c}$$

It is probable that $(\bar{E}_{T/R})_P$ increases with E_{abs} and is smaller than E_{abs}. Thus the internal energy of the nascent product molecules normally increases with E_{abs}. And of course, E_{abs} is controllable over wide limits.

This element of control becomes important when the initial product is capable of further reaction. If E_{abs} is low, the initial product will be stable. But if E_{abs} is high enough, the initial product will be carried on to secondary products.

For example, consider the stepwise breaking of C-Cl bonds in CCl_2F_2.

$$CCl_2F_2 \rightarrow \cdot CClF_2 + \cdot Cl \qquad \Delta H^\circ = 83 \text{ kcal/mole} \tag{5-6}$$

$$\cdot CClF_2 \rightarrow :CF_2 + \cdot Cl \qquad \Delta H^\circ = 49 \text{ kcal/mole} \tag{5-7}$$

Because of the relatively low value of ΔH° for the second step [(5-7)], control of the reaction sequence by controlling E_{abs} should be particularly feasible. In fact, the infrared laser chemistry of CCl_2F_2 has been studied at doses ranging up to 0.5 J/cm^2 where E_{abs} is 25 kcal/mole [11], and also at the higher dose of 1.2 J/cm^2 [70]. At the lower doses there is no evidence for the formation of $:CF_2$; all reaction products are those typical of a free-radical mechanism by way of $\cdot CClF_2$ [11]. At the high dose, however, C_2F_4 is formed, showing that $:CF_2$ is an intermediate [70]. We surmise that at the high dose, E_{abs} was sufficiently large to permit the second step in a substantial fraction of dissociation events.

Besides controlling the internal energy of the primary reaction product and thus its tendency to react further, E_{abs} also controls the *concentration* of the primary reaction product. This can be important for controlling possible chain reactions initiated by the primary product.

For example, B_2H_6 is decomposed conveniently to BH_3 by irradiating at 972.7 cm^{-1}, in the region of the BH_2 wag of $^{11}B_2H_6$ [71].

$$B_2H_6 \rightarrow 2BH_3 \qquad \Delta H^\circ = 37 \text{ kcal/mole} \tag{5-8}$$

The BH_3 molecules then react further; at least one of the processes appears to be a chain reaction yielding the unusual icosaborane (16), $B_{20}H_{16}$.

CW irradiation of 200 torr of B_2H_6 at 1.5 W/cm^2 yielded $B_{20}H_{16}$, while irradiation at 8 W/cm^2 (where the stationary-state concentration of BH_3 presumably is much higher) yielded only products of lower molecular weight [71]. Indeed, laser synthesis of $B_{20}H_{16}$ at 1.5 W/cm^2 was said to be so satisfactory that it may be the method of choice, being superior to conventional synthetic procedures. Icosaborane(16) can be prepared alternatively, in small quantities, by catalytic pyrolysis of $B_{10}H_{14}$ or by high-voltage discharge in $B_{10}H_{14}-H_2$ mixtures between copper electrodes [71]. $B_{20}H_{16}$ is of special interest because of the remarkably low H/B ratio.

Effect of Pressure

Change of pressure has familiar kinetic consequences. First-order processes are favored at low pressure; second- and higher-order processes are favored at higher pressure. However, in infrared laser chemistry kinetic analysis is more complicated, because E_{abs} tends to increase with pressure, and because the molecular energy distribution from which reaction takes place also tends to be affected by the pressure. So far, kinetic analysis in the conventional sense has not yet been possible, at least for the primary steps of laser-induced reactions. Interestingly enough, there is qualitative evidence that laser-induced reactions at high pressures need not be unimolecular. As an example, consider an early study of laser-induced reactions of ethylene in which a CW laser was used to excite the CH_2-wagging mode of ethylene. Products obtained in a 20/80 ethylene/butadiene mixture initially at atmospheric pressure included cyclohexene at 25 W, and cyclopentene and cyclopentadiene at 40 *W* [72]. On pulsed irradiation at megawatt per square centimeter levels, C_2H_4/HCl mixtures yield the addition product C_2H_5Cl [109].

At the opposite pressure extreme, below 1 torr, unimolecular decompositions predominate and, with appropriate choice of infrared wavelength, can be made isotope specific [66].

Use of a Sensitizer

Laser reactions induced by a sensitizer can be controlled in different ways: (1) by choice of sensitizer, giving special consideration to its absorption and energy transfer efficiency, (2) by gas pressure, and (3) by the pressure ratio of reactant to sensitizer. As a case in point, CH_3F has been used to sensitize the decomposition of dioxetane, as shown in (5-9); fluorescence is observed.

$$(CH_3)_2\overset{O-O}{\overset{|\quad\ \ |}{C-C}}(CH_3)_2 \xrightarrow[9.6\mu]{CH_3F} 2CH_3\overset{O}{\overset{\|}{C}}CH_3 + h\nu \qquad (5\text{-}9)$$

E_{abs} under the experimental conditions is relatively small, enough to produce a temperature jump of 160–170°. The sensitized reaction proceeds smoothly, and the product, acetone, is stable [73].

On the other hand, acetone is not stable when a mixture of acetone and SiF_4 is irradiated at moderate doses that are controlled so that E_{abs} can induce *T*-jumps of approximately 500°. Acetone then decomposes largely to ketene and methane. At still higher doses, ethylene, acetylene, and CO also appear [8].

Change of the gas pressure, or of the pressure ratio P_R/P_A of reactant to sensitizer, changes the magnitude of the *T*-jump that can be obtained from E_{abs}. The higher the ratio P_R/P_A, the greater will be the dilution of the energy absorbed by the sensitizer, and the lower will be the effective reaction temperature. For instance, norbornadiene readily undergoes SiF_4-sensitized *retro*-Diels-Alder reaction (5-10).

$$\text{norbornadiene} \xrightarrow[1025\ cm^{-1}]{SiF_4} \text{cyclopentadiene} + HC{\equiv}CH \qquad (5\text{-}10)$$

At a C_7H_8/SiF_4 ratio of 12.5 torr/5.5 torr and a dose of 0.30 J/cm^2, reaction is practically complete in 50 flashes and the product is thermally stable. On the other hand, when cyclopentadiene is irradiated at a C_5H_6/SiF_4 ratio of 10 torr/13 torr at the same dose and number of flashes, 68% of the diene is decomposed [74].

Laser-Driven Explosion

When a continuous laser is used, the amount of radiant energy absorbed per unit time (and hence the reaction rate) will increase with progress of reaction if the product absorbs more strongly than the reactant. In some cases, the effect is so marked that an explosion results. For instance, when ethyl iodide is irradiated at 951 cm^{-1} at 60 W/cm^2, decomposition to ethylene and HI takes place; ethylene absorbs more strongly than the ethyl iodide at this frequency, and the reaction becomes explosive after 2 or 3 min. The explosion manifests itself through a rapid rise in temperature and the formation of I_2 (from the disproportionation of HI) on the walls of the reaction cell. If the laser is tuned to 949 cm^{-1}, where ethylene absorbs less than ethyl iodide, no explosion occurs, although ethylene is formed [75].

Deviations from Thermal Equilibrium

Discussing an insufficiently understood property, such as a system's deviations from thermal equilibrium, in the context of "controllability" may seem premature. However, available facts clearly indicate that controlling such deviations is going to become a potent technique. We discuss two phenomena that have been demonstrated experimentally: (1) For isomerization reactions, the isomer ratio in the infrared photostationary state may be quite different from the thermodynamic ratio; (2) collisional deexcitation by added inert gases may be specifically different for reactant and product and may cause the laser-induced reaction to go to completion.

One of the early gigawatt infrared studies of hydrocarbons involved the *cis-trans* isomerization of 2-butene, shown in (5-11).

$$\underset{\text{cis}}{(H_3C)(H)C{=}C(CH_3)(H)} \xrightleftharpoons[\;]{10.6\,\mu,\; E_{act} = 62\text{ kcal}} \underset{\text{trans}}{(H_3C)(H)C{=}C(H)(CH_3)} \qquad (5\text{-}11)$$

Under the experimental conditions, this process is accompanied by partial decomposition, which has a higher activation energy than the isomerization [76]. When a one-to-one mixture of both isomers is irradiated at 4 torr, a 15% enrichment of the *cis* isomer is observed. At 14 torr, the ratio of the two isomers remains unchanged. Thermodynamic data show that at equilibrium, the *cis/trans* ratio remains less than unity at all temperatures.

The laser-induced isomerization of allene to methylacetylene [given in (5-12)] is conveniently carried out in the presence of SiF_4 sensitizer and is reversible under these conditions [13].

$$H_2C{=}C{=}CH_2 \xrightleftharpoons[]{SiF_4,\ 1025\text{cm}^{-1}} CH_3C{\equiv}CH \qquad (5\text{-}12)$$

Thermodynamic data for this process at 298 °K are $\Delta H° = -1.24$ kcal/mole, $\Delta S° = 1.00$ cal/deg·mole and $\Delta C_p° = +0.40$ cal/deg·mole. Accordingly, the equilibrium constant K is 13.4 at 298 °K, 3.4 at 1000 °K and 2.7 at 2000 °K. A mixture of 20 torr allene/5 torr SiF_4 was irradiated at a dose of 0.73 J/cm^2; results are shown in Fig. 5-6. After about 75 pulses, the system approaches a photostationary state in which the ratio of methylacetylene/allene is about 1.1. On further pulsed irradiation, this ratio increases slowly, reaching 1.6 after 400 pulses.

Inert gases sometimes have a dramatic effect on the yield in laser-driven reactions. For example, the isomerization [given in (5-13)] of

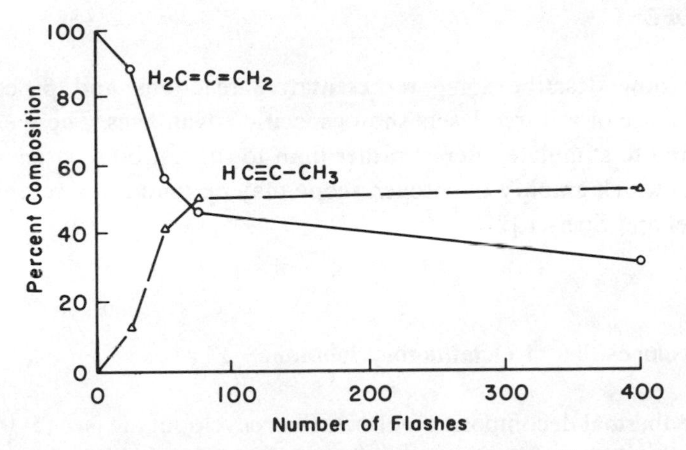

Fig. 5-6 SiF_4-sensitized isomerization of allene to methylacetylene. The percent composition is plotted vs number of laser flashes at 1025 cm^{-1}. Percentages do not add to 100% owing to presence of minor decomposition products. An approximately photostationary state is reached. Based on data in [13].

hexafluorocyclobutene *C* to hexafluorobutadiene *D* has been examined in the presence of helium [68].

$$C \rightleftharpoons D \qquad (5\text{-}13)$$

For this process $\Delta H° = 11.7$ kcal/mole and $\Delta S°$ is 9.50 cal/deg·mole. The equilibrium constant D/C is 0.33 at 1000 °K and 6.3 at 2000 °K. In the laser-induced isomerization of *C* in the absence of helium, the final state is not well-defined but appears to approach 60% *D* at 1 torr. When 16 torr of helium is added to 1 torr of *C*, the CPF decreases by about 50%, but the reaction goes to completion. Evidently, collisional deactivation of *D* by helium atoms is so much more effective than deactivation of *C* that the reverse reaction becomes completely quenched. Note that *D* is thermodynamically the less stable species below 1200 °K, and that for a thermal mechanism the effect of helium would be to reduce the magnitude of the *T*-jump and thus shift the equilibrium toward *C*.

SCOPE

We now describe some representative reactions and processes in which the use of infrared lasers shows specific advantages. These examples are offered to stimulate interest rather than to suggest boundaries. Further examples which amplify the proven scope may be found in a recent review by Kimel and Speiser [77].

Decomposition of Octafluorocyclobutane

The thermal decomposition of octafluorocyclobutane [see (5-14)] takes place smoothly between temperatures of 360 and 560 °C, but side reactions with the wall of the vessel give, in addition to the main product, CO, CO_2, SiF_4, and possibly COF_2 [78].

$$c\text{-}C_4F_8 \longrightarrow 2\ CF_2{=}CF_2 \qquad (5\text{-}14)$$

By contrast, megawatt laser-induced decomposition of this compound at 949 cm^{-1} leads to clean decomposition. The absence of side products was confirmed by infrared spectroscopy and by gas chromatography [16].

The virtual absence of wall effects in gas phase reactions is one of the specific advantages of infrared laser activation, provided that the laser beam is at least slightly smaller then the cross-sectional area of the sample cell. There are two reasons for this: the cell wall remains essentially at room temperature, and the probability that high-energy reactant molecules or labile intermediates will diffuse to the wall without prior reaction is quite small.

Retro-Diels-Alder Reaction of *l*-Limonene

Although the usefulness of the Diels-Alder reaction in organic synthesis has been appreciated for many years, the usefulness of the reverse, *retro*-Diels-Alder reaction has been pointed out only recently [79]. As a matter of historical interest, the *retro*-Diels-Alder reaction [shown in (5-15)] of *l*-limonene *L* to isoprene *I* is one of the early laser-induced organic reaction studies [80].

$$L \longrightarrow 2\ I \qquad (5\text{-}15)$$

According to symmetry selection rules, this reaction is allowed only in the electronic ground state, and the fact that it occurs did confirm at an early date that infrared laser chemistry is electronic ground-state chemistry. This early work was done by direct irradiation of *L* with a 5W CW laser at 943 cm^{-1}. Although *I* was the major product, the chemistry was complex and other products, including benzene, toluene, other substituted aryls, dihydrolimonene, and low-molecular weight compounds, were formed. The chemistry is less complex when *d*-limonene is decomposed in the presence of SiF_4 sensitizer under pulsed megawatt conditions [74].

Aromatic Halogen Substitution

Aromatic substitution is one of the basic reactions of organic chemistry, and it attracted the attention of infrared laser chemists at an early date. For example, the bromination of pentafluorobenzene [shown in (5-16)] is inconveniently slow in the absence of a catalyst.

$$C_6F_5H + Br_2 \longrightarrow C_6F_5Br + HBr \qquad (5\text{-}16)$$

The reaction proceeds at a convenient speed upon irradiation of a gaseous mixture with a 50 W CW laser at 950 cm^{-1}, where C_6F_5H has a strong absorption band [81]. Thus, irradiation of a mixture consisting of 20 torr each of Br_2, C_6F_5H, and Ar for approximately 10 min gives a 45% yield of the desired product with no side reactions occurring. This study is important also because the authors introduced the strategy of changing the added inert gas in order to test whether the laser-induced reaction is photochemical or thermal.

Hexachlorobenzene from Tetrachloroethylene

C_2Cl_4 is known to undergo condensation to C_6Cl_6 at temperatures near 1000°K. However, the process is complicated by side reactions and close temperature control at 975–1000°K is required for good results.

$$3\ Cl_2C{=}CCl_2 \xrightarrow[940\ cm^{-1}]{BCl_3} C_6Cl_6 + 6[Cl] \qquad (5\text{-}17)$$

Cleaner reaction results when the condensation is carried out by infrared laser chemistry with BCl_3 sensitizer [82]. It has been observed that excitation of the B-Cl stretching vibration with a CO_2 laser at 940 cm^{-1} excites a small fraction of the BCl_3 molecules all the way up to the dissociation limit, as evidenced by visible yellow emission [83].

In a typical experiment, the reaction vessel is filled with 150–200 torr of BCl_3 and the 6 W CW laser is turned on, causing the appearance of the yellow BCl_3 luminescence. Tetrachloroethylene is then bled into the cell from a sidearm, and there is a change in the luminescence from yellow to red-orange. The product C_6Cl_6, is collected in another sidearm at 0°C. The yield of C_6Cl_6 is 88% with minor amounts of cyclo-C_5Cl_8, CCl_4, and C_2Cl_6 being formed as by-products. BCl_3 is not truly an inert sensitizer here; it disappears at a rate of about 15% of that of C_2Cl_4. SF_6 does not sensitize the reaction. It is probable that the condensation proceeds by a free-radical mechanism initiated by a BCl_3 dissociation fragment. Interestingly enough, if BCl_3 and C_2Cl_4 are mixed in the dark, a brown solid forms which coats the cell windows.

Infrared Chemiluminescence

In a chemiluminescent reaction, the product is formed in an excited electronic state and subsequently falls to the electronic ground state with the emission of light. When the chemical reaction is a typical thermal one, the light emission takes place continuously until the reactant is exhausted. However, by the use of a pulsed laser, it is possible to cause a suitable reaction to proceed in small steps, one step after each pulse, so that the light is emitted as a series of flashes. For example, the thermal decomposition of dioxetane is chemiluminescent because one of the acetone product molecules is formed in an excited electronic state [84].

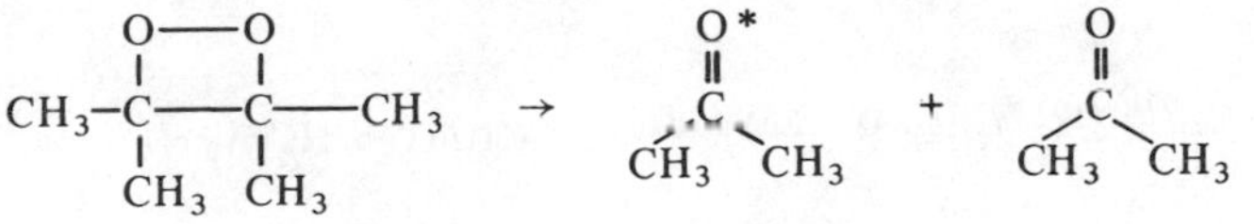

$$\underset{\text{(tetramethyl-1,2-dioxetane: O–O ring on } CH_3-C(CH_3)-C(CH_3)-CH_3)}{} \rightarrow (CH_3)_2C{=}O^* + (CH_3)_2C{=}O \qquad (5\text{-}18)$$

$$(CH_3)_2C{=}O^* \rightarrow (CH_3)_2C{=}O + h\nu(410\ \mathrm{nm}) \qquad (5\text{-}19)$$

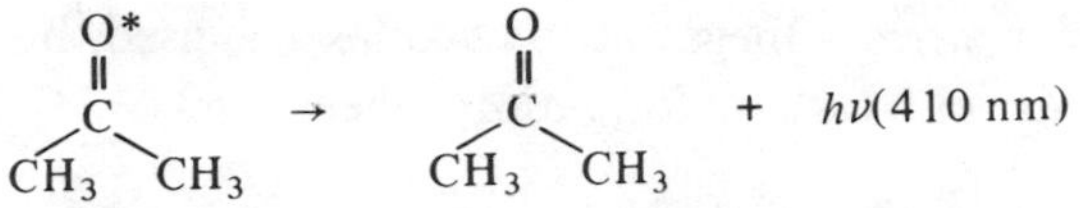

The reaction has been triggered successfully with a pulsed CO_2 laser using CH_3F as sensitizer. At 5 torr CH_3F and 1 torr dioxetane, luminescence reaches a peak 12 μs after the laser flash and decays in about 1 ms [73].

Chloroboration

BCl_3 reacts with benzene at temperatures above 900°K in the presence of palladium catalyst.

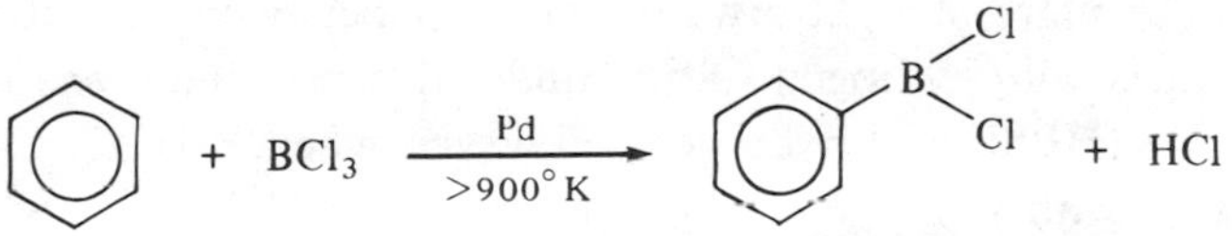

$$C_6H_6 + BCl_3 \xrightarrow[>900^\circ K]{Pd} C_6H_5BCl_2 + HCl \qquad (5\text{-}20)$$

The reaction can be carried out conveniently by pulsed CO_2 laser irradiation at 951 cm^{-1}, where BCl_3 absorbs strongly. For example, a 10 min exposure of a BCl_3-C_6H_6 mixture to 10 μs pulses with a repetition frequency of 1 Hz and a pulse energy of 1 J gives a good yield of $C_6H_5BCl_2$ [1].

BCl_3 reacts with acetylene at 200°C in the presence of Hg_2Cl_2 catalyst according to (5-21).

$$HC{\equiv}CH + BCl_3 \xrightarrow[200\ ^\circ C]{Hg_2Cl_2} ClHC{=}CHBCl_2 \qquad (5\text{-}21)$$

Reaction of BCl_3 with acetylene has also been induced by laser irradiation, but the reaction then yields a substitution product as shown in (5-22).

$$BCl_3 + HC{\equiv}CH \xrightarrow{10.6\ \mu} HCl + HC{\equiv}C{-}BCl_2 \qquad (5\text{-}22)$$

Reaction is clean when a megawatt pulsed laser is used, but is accompanied by pyrolysis and carbon formation when a 50 W CW laser is employed.*

The fact that the laser-induced product is different from the catalytically formed product [see (5-21)] may have practical utility. It may be that the laser-induced reaction mechanism is different, perhaps free-radical vs. ionic. It may also be that the primary laser-induced reaction is like (5-21), followed by HCl elimination.

Reactions of Other Boron Compounds

Because of the complexity frequently found in boron chemistry, infrared lasers offer an independent synthetic approach which in some cases leads to more straightforward chemical behavior. As an example, diborane reacts with hydrogen sulfide under thermal conditions to form a solid polymer $(HBS)_n$ and hydrogen. However, when B_2H_6/H_2S mixtures are irradiated with a 6–7 W continuous laser at 973.3 cm^{-1}, where only B_2H_6 absorbs, a series of reactions ensues and the unusual compounds

*Tar/heating-mantle = carbon/laser.

$(HS)_2BH$ and μ-HSB_2H_5 are formed in isolatable amounts [85]. A probable mechanism is shown in (5-23)–(5-26).

$$B_2H_6 \xrightarrow[973.3\ \text{cm}^{-1}]{\Delta H^\circ = 37\ \text{kcal}} 2\,BH_3 \tag{5-23}$$

$$BH_3 + H_2S \rightarrow HSBH_2 + H_2 \tag{5-24}$$

$$BH_3 + HSBH_2 \longrightarrow \begin{array}{ccccc} H & & H & & H \\ & B & & B & \\ H & & S & & H \\ & & | & & \\ & & H & & \end{array} \tag{5-25}$$

$$H_2S + HSBH_2 \rightarrow (HS)_2BH + H_2 \tag{5-26}$$

There are also side reactions leading to B_5H_9, H_2, and an unidentified solid that may be $(HBS)_n$. The product μ-HSB_2H_5 is of theoretical interest, owing to the existence of a B-S-B bridge, and can be prepared by nonlaser methods only with difficulty.

One reason for the notable differences between laser-induced and thermal reactions of boron compounds is that the laser-induced reactions have a marked tendency to occur from nonequilibrium molecular energy distributions. Evidence to that effect is available both from a study of the decomposition of $H_3B \cdot PF_3$ at low pressure [86], and from alkyl boron chemistry under conditions of practical synthesis [87]. For example, the stepwise reaction sequence of $B(CH_3)_3$ with HBr is shown in (5-27)–(5-29).

$$\underset{970.5\text{cm}^{-1}}{B(CH_3)_3} + HBr \xrightarrow{150\text{–}180\ ^\circ C} B(CH_3)_2Br + CH_4 \tag{5-27}$$

$$\underset{1039.4\text{cm}^{-1}}{B(CH_3)_2Br} + HBr \xrightarrow{>250\ ^\circ C} BCH_3Br_2 + CH_4 \tag{5-28}$$

$$\underset{\substack{970.5\text{cm}^{-1} \\ 1039.4\text{cm}^{-1}}}{BCH_3Br_2} + HBr \xrightarrow{>450\ ^\circ C} BBr_3 + CH_4 \tag{5-29}$$

As shown in the equations, when caused by normal thermal methods the reactions are sharply temperature resolved, so that a heating curve would show three discrete steps.

Laser-induced chemistry is quite different, being determined by which of the substances can absorb the radiant energy. When a mixture of $B(CH_3)_2Br$, $B(CH_3)Br_2$, and HBr is irradiated at 970.5 cm^{-1} (where only $B(CH_3)Br_2$ absorbs), reaction (5-29) takes place but (5-28) does not. Under thermal conditions it would be inconceivable for (5-29) to take place without being accompanied by (5-28). As a control, when $B(CH_3)_3$ and HBr are irradiated at 970.5 cm^{-1}, only $B(CH_3)_2Br$ is formed; there is no further reaction to yield BCH_3Br_2. The data strongly suggest that the absorbed excitation energy remains in the vibrational manifold of the excited species until the moment of reaction.

N_2F_4 Dissociation

One of the remarkable aspects of megawatt laser-induced decompositions is that high concentrations of labile species are produced in a very short time. An early demonstration of this fact is available from the laser-induced decomposition of N_2F_4 [88].

$$N_2F_4 \rightarrow 2 \cdot NF_2 \qquad \Delta H° = 22 \text{ kcal/mole} \qquad (5\text{-}30)$$

The concentration of NF_2 following the 944 cm^{-1} laser flash was monitored by kinetic absorption spectroscopy at 260 nm, where only NF_2 absorbs. Absorption of infrared radiation was measured simultaneously. Since the rate constant for thermal decomposition of N_2F_4 is known, it was possible to compute the NF_2 concentration as a function of time for a thermal reaction mechanism. The computed results are compared with experiment in Fig. 5-7. It is clear that the laser-induced reaction is not thermal.

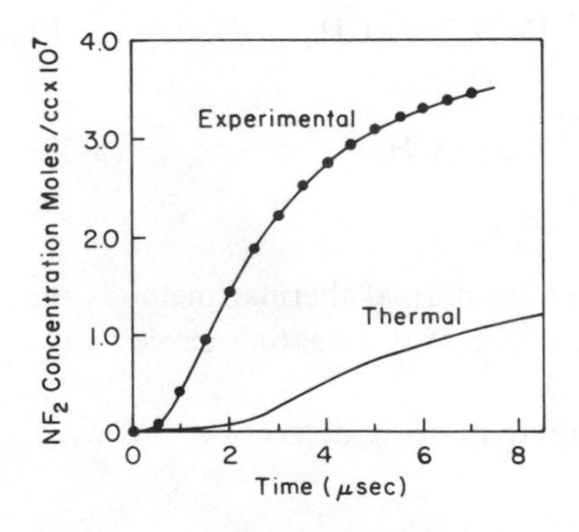

Fig. 5-7 NF_2 concentration vs time after laser flash. Initial N_2F_4 pressure is 22.5 torr. 1.86×10^{-7} moles/cm^3 = 1 torr at 25°C. Note the discrepancy between the experimental curve and the prediction for thermal reaction. Based on data in [88].

Perhaps more important from a practical standpoint are the high concentrations of NF_2 that are obtained. Expressed in terms of partial pressure, the amount of NF_2 formed in Fig. 5-7 is greater than 8 torr, and most of this (about 7 torr) accumulates in 5 μs. It seems safe to predict that the ease and speed with which labile species such as NF_2 can be generated at high concentrations by megawatt infrared laser flashes will open up new opportunities for their use as practical reactants.

Cope Rearrangement of 1, 5-hexadiene

Because of the great difference in the infrared absorption frequencies associated with C-H and C-D bonds, isotopically labeled compounds may sometimes be cleanly rearranged by infrared laser irradiation, so that the isotopic label moves to another site in the molecule. As an example, consider the Cope rearrangement [given in (5-31)] of 1, 5-hexadiene-2,3,3,4,4,5-d_6 [69].

D D D D D D — K ⇌ — D D D D D D

A *B* (5-31)

$K = B/A = 0.71$ (25°), 0.81 (200°)
$=CH_2$ ($=CD_2$) out-of-plane bending frequency:
916 cm^{-1} for A; 714 cm^{-1} for B

When an equilibrium mixture of the two isomers is irradiated with gigawatt pulses at 926 cm^{-1} (where only *A* absorbs), a photostationary state is reached in which the mixture is greatly enriched in *B*. The mole fraction of *B* in the photostationary state increases as the pressure is decreased, and it approaches unity at 5 torr. Rearrangement under these conditions is clean, with little accompanying decomposition. Thus, if one begins with pure *A* isomer, one can produce practically pure *B* isomer. Studies with other deuterium-labeled isomers of 1, 5-hexadiene give consistent results and confirm that the reaction is a genuine concerted Cope rearrangement, rather than dissociation to two allyl radicals followed by recombination.

Effect on Polymer Films

The scope of infrared laser chemistry in the solid state is only beginning to be explored. There are some published reports of significant chemical effects due to exposure to infrared laser radiation, for instance in the hardening of lacquer [89], but it is not clear whether they result from flash-heating or represent genuine photochemistry.

In the case of polymer films, irradiation at an appropriate absorption frequency has been used to obtain more complete polymerization. An example is the condensation polymer, shown in (5-32), sometimes referred to as *PM* polymer owing to its formation from the dianhydride of pyromellitic acid and 4, 4′-diaminodiphenyl ether.

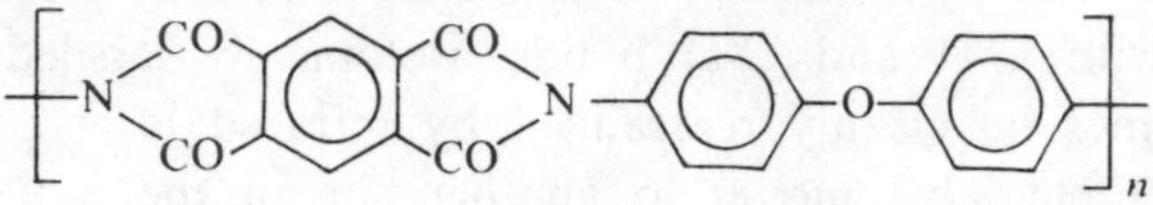

(5-32)

Imidized *PM* polymer

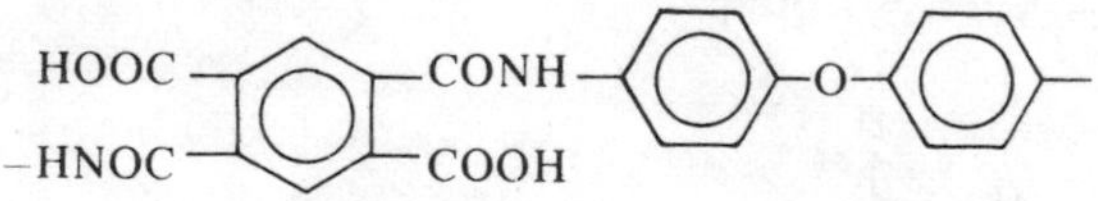

HOOC, CONH, O, –HNOC, COOH (5-33)

Unimidized segment of *PM* polymer

In the formation of the polymer by existing methods of heat treatment, the degree of imidization is only 85%. [Compare (5-32) with (5-33).] When the heat treatment of a 10–20 μ thick film is accompanied by frequent pulsing with 4–8 J/cm^2 infrared radiation tuned to the region of the N-H stretching band at 3260 cm^{-1}, the degree of imidization increases to nearly 100% and improved physical and mechanical properties result [90].

Shift of Chemical Equilibrium

Uranyl nitrate can be extracted from an acidified aqueous solution by shaking with a nonaqueous layer consisting of tributyl phosphate (*TBP*) dissolved in a water-immiscible organic liquid.

$$\underset{945\,cm^{-1}}{UO^{2+}_{aq}} + \underset{1040\,cm^{-1}}{TBP_{org}} \rightleftharpoons UO_2^{2+} \cdot TBP_{org} \qquad (5\text{-}34)$$

By the use of a pulsed 3 MW CO_2 laser, this equilibrium can be shifted so as to favor extraction of uranyl ion into the organic layer [91]. However, the radiation must penetrate to the interface. Measurements were made using a flow system, so as to minimize any temperature rise. When an organic layer consisting of 30% *TBP* in kerosene is floated over the aqueous layer, the distribution coefficient of UO_2^{2+} for extraction into the organic layer goes up by as much as 100% upon irradiation at 944 cm^{-1}, but changes hardly at all upon irradiation at 1044 cm^{-1}. At 944 cm^{-1} the organic layer is transparent to infrared, and the radiation is absorbed by $UO_2^{2+}(aq)$ near the interface. At 1044 cm^{-1}, the radiation is absorbed by *TBP* in the organic layer and does not reach the interface. As a control experiment, when the aqueous layer is floated over an organic layer consisting of 30% *TBP* in carbon tetrachloride, irradiation at 944 cm^{-1} produces no significant effect. In this case the radiation is absorbed by UO_2^{2+} in the bulk of the aqueous phase, and again does not penetrate to the interface [91].

Laser Isotope Separation

Infrared absorption spectra of various isotopic species of the same substance differ slightly, and this property permits some selectivity of excitation. Some representative data are shown in Table 5-1 [93, 94]. As shown there, isotopic frequency shifts tend to be greater for light isotopes (such as $^{12}C/^{13}C$) than for heavy ones (such as $^{79}Br/^{81}Br$).

In laser isotope separation, the laser frequency is tuned so that one isotopic species absorbs more strongly than the others. Because dissociation is sensitive to E_{abs}, the more strongly absorbing species dissociates preferentially and any labile dissociation products are removed by chemical trapping agents. For this method to be successful, it is necessary that collisional exchange of vibrational energy among the various isotopic species be minimized. An obvious way to do this is by working at low pressures. In fact, lowering the pressure generally improves the isotope selectivity of dissociation. However, because the energy-exchanging molecules are isotopic species of the same substance, collisions among them are nearly resonant and energy exchange is correspondingly efficient, so that inconveniently low pressures may be needed.

For light isotopes, isotopic frequency shifts are great enough so that minimizing the importance of vibrational energy exchange and of its

Table 5-1 Typical Isotopic Frequency Shifts

Substance	Frequency
$^{12}CF_2Cl_2$	922, 1099, 1162
$^{13}CF_2Cl_2$	876, 1077, 1131
$^{28}SiF_4$	389.4, 1031.8, 1190.4, 1825.7
$^{29}SiF_4$	387.8, 1022.9, 1188.4, 1816.9
$^{32}SF_6$	614.5, 947.9
$^{34}SF_6$	611.2, 930.5
$^{79}BrCD_3$	576.72 (C-Br stretch)
$^{81}BrCD_3$	575.51
$^{238}UF_6$	667(ν_1), 624(ν_3)
$^{235}UF_6$	(668.4)[a], (624.7)[a]

[a] Estimated; ν_1 is infrared inactive.

spoiling of isotope-selective dissociation are the principal problems. For heavy isotopes, additional problems arise that are considered later.

There are basically two ways in which a chemist can minimize vibrational energy exchange and its consequences: by appropriate choice of the substance and by control of the dissociation. In choosing the substance, a useful concept is to regard the more strongly absorbing isotopic species as a sensitizer for vibrational excitation of the other species. It then follows that those substances will work best whose excitation is not easily sensitized. For example, data in Table 4-3 show that energy transfer from SiF_4 to CCl_2F_2 is relatively inefficient. It appears, therefore, that excitation of CCl_2F_2 is not easily sensitized. This property makes CCl_2F_2 a reasonable candidate for attempting carbon or chlorine laser-isotope separation.

As a matter of fact, experiments done with CCl_2F_2 lead to good carbon isotopic enrichment [92, 93]. In a typical experiment, 1.1 torr of CCl_2F_2 and 3.32 torr of dimethylethylene was irradiated at an intensity of 10 MW/cm^2 at 929 cm^{-1}, where $^{12}CCl_2F_2$ is excited. It was found, after several thousand flashes of irradiation, that the unreacted CCl_2F_2 was enriched in $^{13}CCl_2F_2$ by the very substantial factor of 3.8. The fluorine-containing products (C_2F_4, $CH_2{=}CF_2$ and $CHClF_2$) were correspondingly depleted in ^{13}C by factors ranging from 1.4 to 6 [92, 93]. The trapping reaction is probably the reaction of $:CF_2$ with $(CH_3)_2C{=}CH_2$ to form 1, 1-difluoro-2, 2-dimethylcyclopropane, which is unstable under the reaction conditions and goes on to the detected products [93].

The success of using CCl_2F_2 as a substrate for carbon isotope enrichment may be due in part to the nature of the decomposition reaction. As stated earlier [see (5-6) and (5-7)], decomposition probably takes place in two consecutive steps, depending on the magnitude of E_{abs}. Although the evidence for this needs to be strengthened, the *principle* of using a two-step decomposition process for laser isotope separation is sound. If conditions are chosen so that the relative importance of the second step depends markedly on the magnitude of E_{abs}, and if the trapping agent reacts better with the product of the second step than with that of the first, the dissociation chemistry will enhance the selectivity resulting even from small isotopic differences in E_{abs}.

Laser isotope separation of molecules containing heavy isotopes is more difficult than that of light isotopes because one encounters the following vicious circle. (1) If one works at pressures above 1 torr, collisional transfer of vibrational energy effectively spoils the isotopic selectivity of excitation. (2) If one works at low pressure, the dose extinction coefficient e_A suffers [see discussion of (3-5) and (3-6), pages 27–29]. Thus high radiant intensities are needed. (3) If high intensities are used, coherent power broadening of the absorption lines becomes greater than the isotopic frequency shift, and the potential for isotopically selective excitation is lost.

The magnitude of the coherent power broadening is given by (5-35), where ω_R is the Rabi frequency, and c is the speed of light [35].

$$(\delta\nu)_{\text{power broadening}} = \frac{\omega_R}{2\pi c}(\text{cm}^{-1}) \qquad (5\text{-}35)$$

The Rabi frequency is proportional to the amplitude E_0 of the electric vector of the coherent radiation and to the transition moment $|\mu|$ of the infrared absorption band, as in (5-36).

$$\omega_R = 2\pi|\mu|\frac{E_0}{h} \qquad (5\text{-}36)$$

For strong infrared absorption bands, $|\mu|$ is in the range 0.1–0.5 D. E_0 is related to the (r.m.s.) intensity by $E_0 = 27.5\ I^{1/2}$ V/cm when I is expressed in W/cm^2. Thus, when $I = 1$ GW/cm^2, the power broadening is typically 2–10 cm^{-1}, as compared to isotopic frequency shifts of approximately 1 cm^{-1} for $^{238}UF_6$ vs $^{235}UF_6$.

To break this vicious circle, a two-photon excitation technique has been used [95]. To minimize the effects of power broadening, one uses a carefully tuned, relatively weak, megawatt-level laser to excite the desired isotopic molecules into the quasi-continuum (Fig. 3-9). At the same time, one uses a more powerful laser tuned to a lower frequency in order to lift the molecules within the quasi-continuum to the dissociation limit.

For instance, 0.3 torr of OsO_4 and 2.0 torr of the trapping agent OCS was irradiated simultaneously with two laser frequencies, both of which are in the region of the 960 cm^{-1} OsO_4 absorption band. The ν_1 frequency was 956.2 cm^{-1}, at the edge of the Q branch; the ν_2 frequency was 944 cm^{-1}, far into the P branch. Both laser beams were unfocused, with intensities in the MW/cm^2 range. The resulting isotopic enrichment of the unreacted OsO_4 was as follows:

$^{192}Os/^{190}Os$	1.11
$^{192}Os/^{189}Os$	1.13
$^{192}Os/^{188}Os$	1.24
$^{192}Os/^{187}Os$	1.48
$^{192}Os/^{186}Os$	1.58

By comparison, single-frequency irradiation of the same mixture at 958 cm^{-1} with an intensity of 100 MW/cm^2 gave no significant isotopic enrichment [95]. Evidently, in the absence of pressure broadening, the ^{186}Os species is lifted more efficiently from the ground state into low-lying excited levels by 956.2 cm^{-1} radiation than are the other isotopic species. Once the molecules are there, the 944 cm^{-1} radiation is effective at lifting them to the dissociation limit.

Because the two-frequency isotopic separation process does not require focused laser beams, larger gas volumes can be irradiated and scaling up to an industrial process can be contemplated. The technique of irradiation at two discrete laser frequencies, the practical benefits of which are here so clearly demonstrated, should also be applicable to the simultaneous excitation of remote centers in extended molecules, and should provide new knowledge, not only concerning infrared laser-induced reactions, but also concerning intramolecular energy transfer.

Appendix I

Terminology, Units, and Constants

PHOTOPHYSICAL TERMINOLOGY

The American Institute of Physics Handbook [96] gives the following definitions:

> *Radiant flux* or *radiant power* is the rate at which energy is radiated; it is commonly expressed in watts.
> *Radiant energy* is the time integral of radiant flux; it is consistently expressed in joules.

In working with laser beams, one needs specific terms to denote radiant flux and radiant energy *per unit area*, the area being measured normal to the direction of propagation of the laser beam. Official terms for these quantities apparently do not exist. Prompted by related usage and precedent, we use the terms *beam intensity* or simply *intensity*, (in watts per square centimeter) and *dose* (in joules per square centimeter), respectively.

Beam intensity is an obvious variant of *radiant intensity*, the flux per unit solid angle emanating from a point source. In colloquial usage, the term "intensity" may refer either to the strength of a source or to that of a beam. The sense-perception of "intensity" is complicated but, roughly speaking, measures the flux entering the pupil of the human eye and thus approximately the beam intensity. In photochemical studies, the rate of photoactivation is basically proportional to the radiant energy density. However, it is often said to be proportional simply to the "intensity," whose meaning in this context is that of beam intensity, as shown in (A-1).

$$\text{Beam intensity } (\text{W/cm}^2) = \text{speed of light } (\text{cm/s}) \times \text{radiant energy density } (\text{J/cm}^3) \quad (A\text{-}1)$$

According to electromagnetic theory, root-mean-square beam intensity (I) of a coherent laser beam is related to the amplitude E_0 of the electric field vector according to (A-2), where Z_0 is the impedance of free space. In practical units, Z_0 is 377 ohms.

$$I\,(\text{W/cm}^2) = \frac{E_0^2}{2Z_0}(\text{V}^2/\text{cm}^2\text{ohm}) \quad (A\text{-}2)$$

The term *dose* has long been used in radiation-related areas such as radiology, radiation chemistry, and health physics to denote the amount of radiation entering, or falling on, an object. A variety of practical units are in use, depending on the nature and convenience of the problem. In infrared laser chemistry it is convenient to let *dose* denote the amount of radiant energy passing through unit area in a plane normal to the beam direction. Convenient units are joules per square centimeter.

Although the term "dose" is concise, familiar, and sanctioned by precedent, it should be emphasized that its use in this context has not been generally accepted. The term *energy fluence* has also been proposed and seems to be gaining in favor among laser physicists.

UNITS

Scientific data in this volume are expressed in practical units that are popular among chemists and approved for use in the journals published by the American Chemical Society. Thus concentrations are in moles per liter, gas pressures in torr or atmospheres, and activation energies in kilocalories per mole. E_{abs} is expressed in kilocalories per mole because it is convenient for E_{abs} and E_{act} to be expressed in the same units. Other quantities such as wavenumber, beam intensity, dose, and electric field intensity are expressed in practical cgs units.

CONVERSION FACTORS

Energy-related units:

$$1\ \text{cm}^{-1} = 2.858\ \text{cal/mole}$$
$$= 11.96\ \text{J/mole}$$
$$= 1.240 \times 10^{-4}\ \text{eV}$$
$$= 1.986 \times 10^{-16}\ \text{erg/molecule}$$

$$4.185\ \text{J} = 1\ \text{cal}$$

$$1\ \text{J} = 10^{7}\ \text{erg}$$
$$= 0.00987\ \text{liter atm}$$

Pressure-related units:

$$1\ \text{torr} = 1/760\ \text{atm}$$
$$= 1.333\ \text{millibar}$$
$$= 5.38 \times 10^{-5}\ \text{mole/liter at } 298.16^{\circ}\text{K}$$

GENERAL PHYSICAL CONSTANTS

c	Speed of light	2.9979×10^{10} cm/s
e	Electronic charge	9.649×10^{4} coulomb/mole
		1.602×10^{-19} coulomb/electron
h	Planck's constant	6.6255×10^{-34} J·s
k	Boltzmann constant	1.3805×10^{-23} J/°K
N_0	Avogadro's number	6.0226×10^{23} molecules/mole
R	Gas constant per mole	1.9867 cal/mole °K
		8.314 J/mole °K
		8.205×10^{-2} liter atm/mole °K
Z_0	Impedance of free space	376.7 ohm

Appendix II

Infrared Laser Emission Lines

Number	Emitting Gas	Range (obs., cm^{-1})	References
1	CO_2	908–1092	97,99
2	CO	1966–1632	97,98
3	N_2O	906–961	99
4	HC1	2698–2482	97,100
5	DC1	1982–1832	97,100
6	HF	3834–3233	97,99
7	DF	2863–2439	97,99

CO_2 LASER

Branch (Number)	Wavenumber (cm^{-1})	Approximate Strength	Branch (Number)	Wavenumber (cm^{-1})	Approximate Strength
		00°1–10°0 Transition			
P(56)	907.78	0.3	R(4)	964.77	0.3
P(54)	910.02	0.3	R(6)	966.25	0.65
P(52)	912.23	0.3	R(8)	967.71	0.75
P(50)	914.42	0.3	R(10)	969.14	0.8
P(48)	916.58	0.55	R(12)	970.55	0.85
P(46)	918.72	0.55	R(14)	971.93	0.95
P(44)	920.83	0.6	R(16)	973.29	0.95
P(42)	922.92	0.7	R(18)	974.62	1.0
P(40)	924.97	0.75	R(20)	975.93	1.0

Branch (Number)	Wavenumber (cm^{-1})	Approximate Strength	Branch (Number)	Wavenumber (cm^{-1})	Approximate Strength
00°1–10°0 Transition (continued)					
P(38)	927.01	0.85	R(22)	977.21	1.0
P(36)	929.02	0.85	R(24)	978.47	1.0
P(34)	931.00	0.95	R(26)	979.71	0.95
P(32)	932.96	0.95	R(28)	980.91	0.9
P(30)	934.90	1.0	R(30)	982.10	0.9
P(28)	936.80	1.0	R(32)	983.25	0.85
P(26)	938.69	1.0	R(34)	984.38	0.85
P(24)	940.55	1.0	R(36)	985.49	0.8
P(22)	942.38	1.0	R(38)	986.57	0.75
P(20)	944.19	1.0	R(40)	987.62	0.75
P(18)	945.98	1.0	R(42)	988.65	0.6
P(16)	947.74	1.0	R(44)	989.65	0.45
P(14)	949.48	1.0	R(46)	990.62	0.35
P(12)	951.19	1.0	R(48)	991.57	0.25
P(10)	952.88	0.95			
P(8)	954.55	0.85	R(50)	992.49	—
P(6)	956.19	0.75			
P(4)	957.80	0.55			
00°1–02°0 Transition					
P(50)	1016.72	0.15	R(4)	1067.54	0.25
P(48)	1018.90	0.25	R(6)	1069.01	0.4
P(46)	1021.06	0.4	R(8)	1070.46	0.5
P(44)	1023.19	0.55	R(10)	1071.88	0.6
P(42)	1025.30	0.55	R(12)	1073.28	0.6
P(40)	1027.38	0.65	R(14)	1074.65	0.6
P(38)	1029.44	0.65	R(16)	1075.99	0.6
P(36)	1031.48	0.7	R(18)	1077.30	0.6
P(34)	1033.49	0.7	R(20)	1078.59	0.6
P(32)	1035.47	0.75	R(22)	1079.85	0.55
P(30)	1037.43	0.75	R(24)	1081.09	0.55
P(28)	1039.37	0.75	R(26)	1082.30	0.55
P(26)	1041.28	0.8	R(28)	1083.48	0.5
P(24)	1043.16	0.8	R(30)	1084.64	0.5
P(22)	1045.02	0.8	R(32)	1085.77	0.5
P(20)	1046.85	0.75	R(34)	1086.87	0.5
P(18)	1048.66	0.75	R(36)	1087.95	0.4
P(16)	1050.44	0.7	R(38)	1089.00	0.35
P(14)	1052.20	0.7	R(40)	1090.03	0.35
P(12)	1053.92	0.7	R(42)	1091.03	0.25
P(10)	1055.63	0.7	R(44)	1092.01	0.2
P(8)	1057.30	0.6			
P(6)	1058.95	0.35			
P(4)	1060.57	0.15			

CO LASER

Vibrational Transition	Rotational Transition	Wavenumber Range (cm^{-1})
5→4	P(18)→P(27)	1966–1926
6→5	P(16)→P(28)	1949–1896
7→6	P(15)→P(28)	1927–1871
8→7	P(14)→P(28)	1906–1855
9→8	P(15)→P(28)	1876–1821
10→9	P(14)→P(27)	1855–1800
11→10	P(13)→P(25)	1834–1784
12→11	P(17)→P(25)	1792–1759
13→12	P(15)→P(24)	1775–1739
14→13	P(15)→P(25)	1750–1710
15→14	P(14)→P(24)	1729–1689
16→15	P(16)→P(22)	1697–1673
17→16	P(16)→P(20)	1672–1656
18→17	P(15)→P(20)	1651–1632

N_2O LASER

Vibrational Transition	Rotational Transition	Wavenumber Range (cm^{-1})
00°1→10°0	P(37)→P(5)	905.7–934.6
	R(3)→R(27)	942.1–960.8

HCl CHEMICAL LASER

Vibrational Transition	Rotational Transition	Wavenumber Range (cm^{-1})
2→1	P(4)→P(10)	2698–2554
3→2	P(4)→P(9)	2597–2482

DCl CHEMICAL LASER

Vibrational Transition	Rotational Transition	Wavenumber Range (cm^{-1})
2→1	P(5)→P(9)	1982–1935
3→2	P(4)→P(11)	1941–1859
4→3	P(5)→P(9)	1878–1832

HF CHEMICAL LASER

Branch (Number)	Wavenumber (cm^{-1})	Approximate Strength	Branch (Number)	Wavenumber (cm^{-1})	Approximate Strength
	1→0 Transition			2→1 Transition	
P(3)	3834	0.05	P(3)	3666.4	0.2
P(4)	3788	0.3	P(4)	3622.7	0.2
P(5)	3741	—	P(5)	3577.5	0.45
P(6)	3693.5	0.4	P(6)	3531.3	0.6
P(7)	3644.2	1.0	P(7)	3483.6	0.75
P(8)	3593.8	0.5	P(8)	3435.2	0.75
P(9)	3542.2	0.35	P(9)	3385.3	0.25
	3→2 Transition				
P(3)	3503.8	—			
P(4)	3461.5	0.15			
P(5)	3418.2	0.2			
P(6)	3373.5	0.1			
P(7)	3327.7	0.05			
P(8)	3280.6	—			
P(9)	3233	0.05			

DF CHEMICAL LASER

Branch (Number)	Wavenumber (cm^{-1})	Approximate Strength	Branch (Number)	Wavenumber (cm^{-1})	Approximate Strength
	1→0 Transition			2→1 Transition	
P(2)	2863	0.03	P(3)	2750.1	0.3
P(3)	2840	0.15	P(4)	2727.4	0.3
P(4)	2817	0.3	P(5)	2704.0	0.45
P(5)	2793	0.4	P(6)	2680.3	0.7
P(6)	2769	0.6	P(7)	2656.0	0.5
P(7)	2743	0.8	P(8)	2631.1	1.0
P(8)	2718	0.8	P(9)	2605.9	0.9
P(9)	2693	0.65	P(10)	2580.2	0.65
P(10)	2666	0.5	P(11)	2554.0	0.4
P(11)	2639	0.4	P(12)	2527.5	0.2
P(12)	2611.1	0.3	P(13)	2500.3	0.15
	3→2 Transition			4→3 Transition	
P(2)	2684	0.25	P(6)	2509.9	0.07
P(3)	2662.2	0.25	P(7)	2486.3	0.03
P(4)	2640.0	—	P(8)	2463	0.03
P(5)	2617.4	0.3	P(9)	2439	0.03
P(6)	2594.2	0.55			
P(7)	2570.5	0.6			
P(8)	2546.4	0.6			
P(9)	2521.8	0.45			
P(10)	2496.6	0.35			
P(11)	2471.3	0.15			
P(12)	2445.3	0.1			

Appendix III

Infrared Optical Materials

Abbreviations:

tr	Transmission range to 10% points (sample thickness)
refl	Reflectance
h	Hardness, moh scale: 1 (soft) to 10 (very hard)
hygr	Hygroscopic
n	Refractive index

Companies that grow and fabricate infrared optical materials are listed in the *Laser Focus Buyer's Guide* (1976), p. 150. References: 102–107.

Alumina (Al_2O_3)	See corundum, ruby, sapphire.
Arsenic trisulfide (As_2S_3)	tr, 0.6–12 μ (5 mm); refl, 25%@5 μ; h, 2; nontoxic; inexpensive.
Barium fluoride (BaF_2)	tr, 0.16–15 μ (10 mm); refl, 5%@10 μ; h, 3; inexpensive; sensitive to thermal shock; can be used in pressure cells; vacuum seals can be made with AgCl or suitable epoxy.
Cadmium selenide (CdSe)	tr, 0.75–25 μ (1.67 mm); n, 2.43@1.1 μ, birefringent.
Cadmium sulfide (CdS)	tr, 0.5–16 μ (2 mm); n, 2.3@1.1 μ, birefringent.
Cadmium telluride (CdTe)	tr, 0.9–33 μ (2 mm) n, 2.67@10 μ.

Calcite	See calcium carbonate.
Calcium carbonate ($CaCO_3$)	tr, 0.25–3 μ; refl, 5%@2 μ; h, 3.
Calcium fluoride (CaF_2)	tr, 0.15–12 μ (2 mm); refl, 4%@5 μ; h, 4; sensitive to thermal shock; can be used in pressure cells; vacuum seals can be made with AgCl or suitable epoxy.
Carbon (C)	See Diamond.
Cesium bromide (CsBr)	tr, 0.2–45 μ (1 cm); refl, 8%@10 μ; h, 2; hygr; resistant to thermal shock.
Cesium chloride (CsCl)	tr, 0.19–30 μ; refl, 11%@10 μ; h, 2; hygr.
Cesium fluoride (CsF)	tr, 0.27–> 15 μ; refl, 8%@10 μ; h, 2; very hygr;
Cesium iodide (CsI)	tr, 0.25–60 μ (5 mm); refl, 9%@10 μ; h, 1.5; hygr; resistant to thermal shock.
Cuprous chloride (CuCl)	tr, 0.4–19 μ (9.1 mm); n, 1.92@5 μ; h, 2.5.
Corundum (α-Al_2O_3)	tr, 0.17–6.5 μ; n, 1.76@0.69 μ, birefringent; h, 9.
Diamond (C)	tr, 0.23–100+ μ (2 mm); n, 2.41@0.66 μ; h, 10; durable except at high temperatures under oxidizing conditions.
Gallium arsenide (GaAs)	tr, 1-15 μ (0.5 mm); n, 3.31@4 μ; expensive; one of the best materials for high power CO_2 laser components; good thermal and mechanical properties.
Germanium (Ge)	tr, 1.8–26 μ and 38–70 μ (1.5 mm); refl, 39% @ 10 μ; inexpensive without coatings; useful at low power densities; high thermal conductivity.
Irtran 1	See Magnesium fluoride.
Irtran 2	See Zinc sulfide.
Irtran 3	See Calcium fluoride.
Irtran 4	See Zinc selenide.
Irtran 5	See Magnesium oxide.
Irtran 6	See Cadmium telluride.
KRS-5	See Thallium bromoiodide.

KRS-6	See Thallium bromochloride.
Lead fluoride (β-PbF_2)	tr, 0.29–11.6 μ; refl, 15%@10 μ; h, 3.
Lithium fluoride (LiF)	tr, 0.1–7 μ (10 mm); refl, 4%@5 μ; h, 3; sensitive to thermal shock; vacuum seals can be made with AgCl or suitable epoxy.
Magnesium fluoride (MgF_2)	tr, 0.11–8 μ (1 mm); refl, 4%@5 μ; h, 5.
Magnesium oxide (MgO)	tr, 0.25–9 μ (1 mm); refl, 10%@5 μ; h, 5.5.
Potassium bromide (KBr)	tr, 0.2–35 μ (10 mm); refl, 8%@10 μ; h, 2; inexpensive.
Potassium chloride (KCl)	tr, 0.18–25 μ (10 mm); refl, 4%@10 μ; h, 2; inexpensive.
Potassium fluoride (KF)	tr, 0.16–15 μ; refl, 5%@10 μ; h, 2.5, hygr.
Potassium iodide (KI)	tr, 0.25–40 μ (6 mm); n, 1.63@5 μ, h, 1.5; hygr.
Quartz (SiO_2)	tr, 0.1–4 μ and 43–50+ μ (1 cm); refl, 3%@2 μ; h, 7.
Ruby (Al_2O_3+0.5%Cr)	tr, 1–6.5 μ (6.1 mm); refl, 8%@3 μ; h, 9.
Rutile	See Titanium dioxide.
Sapphire (Al_2O_3)	tr, 0.2–6.5 μ (3 mm); refl, 10%@2 μ; h, 9.
Selenium (Se)	tr, 1–20 μ (1.69 mm); n, 2.5@10 μ.
Silica (SiO_2)	tr, 0.15–4.5 μ (3 mm); n, 1.52@2 μ; h, 5.5.
Silicon (Si)	tr, 1.2–50+ μ (2.5 mm); refl, 30%@10 μ; h, 7.
Silver bromide (AgBr)	tr, 0.45–42 μ (1 mm); n, 2.23@0.7 μ; h, 2; cannot be used in pressure cells; cannot be used above 200°C; cold-flows.
Silver chloride (AgCl)	tr, 0.4–34 μ (0.5 mm); refl, 16%@10 μ; h, 1.5; cannot be used in pressure cells; cannot be used above 200°C; cold-flows.
Sodium bromide (NaBr)	tr, 0.2–23 μ; refl, 12%@10 μ; h, 2.5, hygr.
Sodium chloride (NaCl)	tr, 0.2–20 μ (5 mm); refl, 4%@10 μ; h, 2.5; sensitive to thermal shock.

Sodium fluoride (NaF)	tr, 0.13–15 μ (2.16 mm); refl, 2% @10 μ; h, 3.5.
Sodium iodide (NaI)	tr, 0.25–25 μ; refl, 15% @10 μ; h, 2; hygr.
Sodium nitrate ($NaNO_3$)	tr, 0.35–3 μ; h, 2.5; hygr; birefringent.
Strontium barium niobate ($Sr_{0.5}Ba_{0.75}Nb_2O_6$)	tr, 0.33–7 μ (3 mm); n, 2.22 @1 μ; birefringent.
Strontium fluoride (SrF_2)	tr, 0.13–14 μ (3 mm); n, 1.44 @0.5 μ; h, 3.
T-12	tr, 0.5–12 μ (4 mm); refl, 5% @5 μ; vacuum seals can be made with AgCl or suitable epoxy.
Thallium bromide (TlBr)	tr, 0.45–48 μ (1.65 mm); refl, 22% @10 μ; h, 2; toxic.
Thallium bromochloride (Tl_2BrCl)	tr, 0.4–35 μ (1.65 mm); refl, 22% @10 μ, h, 2.5; toxic.
Thallium bromoiodide (Tl_2BrI)	tr, 0.6–50 μ (2 mm); refl, 20% @10 μ, h, 2; cold-flows; toxic.
Thallium chloride (TlCl)	tr, 0.4–34 μ (1.65 mm); refl, 20% @10 μ; h, 2.5; toxic.
Titanium dioxide (TiO_2)	tr, 0.43–6.2 μ (5 mm); refl, 18% @2 μ; h, 6.
Zinc selenide (ZnSe)	tr, 0.5–22 μ (2 mm); refl, 23% @10 μ; expensive; excellent material for use in transmitting high-power laser light.
Zinc sulfide (ZnS)	tr, 0.6–16 μ (1 mm); refl, 21% @10 μ; h. 4.

References

1. N. V. Karlov, *Appl. Opt.* **13** (1974) 301 [reviews early work].
2. R. V. Ambartzumian and V. S. Letokhov, *Acc. Chem. Res.* **10** (1977) 61 [reviews current work].
3. J. L. Lyman and S. D. Rockwood, *J. Appl. Phys.* **47** (1976) 595.
4. J. L. Lyman, R. J. Jensen, J. Rink, C. P. Robinson, and S. D. Rockwood, *Appl. Phys. Lett.* **27** (1975) 86.
5. C. K. Patel, "High Power Lasers," *Sci. Am.*, August 1968, pp. 22 ff.
6. K. L. Kompa, "Chemical Lasers," *Topics in Current Chemistry*, Vol. 37, Springer-Verlag, New York, 1973.
7. K. A. Kovaly (Ed.), "Inside R & D," Technical Insights, Inc. New York, **6**(11) (1977) 1.
8. K. J. Olszyna, E. Grunwald, P. M. Keehn, and S. P. Anderson, *Tetrahedron Lett.* **19** (1977) 1609.
9. D. F. Dever and E. Grunwald, *J. Am. Chem. Soc.* **98** (1976) 5055.
10. E. Grunwald, K. J. Olszyna, D. F. Dever, and B. Knishkowy, *J. Am. Chem. Soc.* **99** (1977) 6515.
11. G. A. Hill, E. Grunwald, and P. Keehn, *J. Am. Chem. Soc.* **99** (1977) 6521.
12. R. Walsh, *J. Chem. Soc. Faraday Trans.* **72** (1976) 9.
13. P. Keehn and C. Cheng, *J. Am. Chem. Soc.* **99** (1977) 5808.
14. W. M. Shaub and S. H. Bauer, *Int. J. Chem. Kinetics* **12** (1975) 509.
15. W. C. Danan, W. D. Munslow, and P. W. Setzer, *J. Am. Chem. Soc.* **99** (1977) 6961.
16. J. M. Preses, R. E. Watson, Jr., and G. W. Flynn, *Chem Phys. Lett.* **46** (1977) 69.
17. C. W. Mathews, *Can J. Phys.* **51** (1973) 1281.
18. E. Grunwald, C. M. Lonzetta, and K. J. Olszyna, *Am. Chem. Soc. Symp. Ser.* **56** (1977) 107.
19. N. R. Isenor, *Can. J. Phys.* **51** (1973) 1281.
20. V. Ambartzumian, N. V. Chekalin, V. A. Doljikov, V. S. Letokhov, and V. N. Lokhuan, *J. Photochem.* **6** (1976/77) 55.
21. G. Francis, *Ionization Phenomena in Gases*, Academic Press, New York, 1960, Chapter 4.
22. N. R. Isenor and M. C. Richardson, *Appl. Phys. Lett.* **18** (1971) 224.
23. S. L. Chin, *Phys. Lett.* **61A** (1977) 311.

24. J. Cooper, *Rev. Sci. Inst.* **41** (1970) 565.
25. E. D. West and D. A. Jennings, *Rev. Sci. Inst.* **41** (1970) 565.
26. A. G. Bell, *Proc. Am. Assoc. Adv. Sci.* **29** (1880) 115.
27. W. R. Harshbarger and M. S. Robin, *Acc. Chem. Res.* **6** (1973) 329.
28. J. C. Black, E. Yablonovich, N. Bloembergen, and S. Mukamel, *Phys. Rev. Lett.* **38** (1977) 1131.
29. T. F. Deutsch, *Opt. Lett.* **1** (1977) 25.
30. V. N. Bagratashvili, I. N. Knyazev, V. S. Letokhov, and V. V. Lobko, *Opt. Comm.* **18** (1976) 525.
31. N. W. Ressler and R. W. Crain, *Appl. Phys. Lett.* **20** (1972) 42.
32. C. Steel and V. Sarakhov, private communication, August 1977.
33. R. N. Schwartz, Z. I. Slawsky, and K. F. Herzfeld, *J. Chem. Phys.* **23** (1955) 1118.
34. S. Mukamel and J. Jortner, *J. Chem. Phys.* **65** (1976) 3735.
35. N. Bloembergen, *Opt. Comm.* **15** (1975) 416.
36. R. P. Feynman, *Phys Rev.* (Series 2) **84** (1951) 108; **80** (1950) 451.
37. D. Frankel, J. T. Steinfeld, R. D. Sharma, and L. Poulsen, *Chem. Phys. Lett.* **28** (1974) 485.
38. K. F. Herzfeld and T. A. Litovitz, *Absorption and Dispersion of Ultrasonic Waves*, Academic Press, New York, 1959.
39. E. Weitz and G. Flynn, *Ann. Rev. Phys. Chem.* **25** (1974) 275.
40. J. D. Lambert and R. Salter, *Proc. R. Soc.* **A253** (1959) 277.
41. J. M. Preses, Ph.D. dissertation, Columbia University, 1975, p. 64.
42. T. D. Rossing and S. Legvold, *J. Chem. Phys.* **23** (1955) 1118.
43. F. London, *Z. Phys. Chem.* (*Leipzig*) **B11** (1930) 222.
44. D. C. Tandy and B. S. Rabinovitch, *Chem. Rev.* **77** (1977) 369.
45. J. E. Leffler and E. Grunwald, *Rates and Chemical Equilibria of Organic Reactions*, John Wiley and Sons, New York, 1963, p. 100.
46. E. K. Plyler and N. Aquista, *J. Res. N. B. S.* **48** (1952) 92.
47. E. K. Plyler and W. S. Benedict, *J. Res. N. B. S.* **47** (1951) 202.
48. D. Gutman, W. Braun, and W. Tsang, *J. Chem. Phys.* **67** (1977) 4291.
49. C. E. Treanor, J. W. Rich, and R. G. Rehm, *J. Chem. Phys.* **48** (1968) 1798, Appendix.
50. R. M. Osgood, P. B. Sackett, and A. Javan, *J. Chem. Phys.* **60** (1974) 1464.
51. B. L. Earl and A. M. Ronn, *Chem. Phys. Lett.* **39** (1976) 95.
52. J. D. Teare, R. L. Taylor, and C. W. von Rosenberg, *Nature*, **225** (1970) 240.
53. F. I. Tanczos, *J. Chem. Phys.* **25** (1956) 439.
54. I. Shamah and G. Flynn, *J. Am. Chem. Soc.* **99** (1977) 3191.
55. R. E. McNair, S. F. Fulghum, G. W. Flynn, and M. S. Field, *Chem. Phys. Lett.* **48** (1977) 241.
56. S. Mukamel and J. Ross, *J. Chem. Phys.* **66** (1977) 35.
57. D. L. Bunker, *Acc. Chem. Res.* **7** (1974) 195.
58. W. H. Hase, "Dynamics of Unimolecular Reactions," *Dynamics of Molecular Collisions, Part B*, W. Miller, Ed. Plenum Press, New York, 1976.
59. P. J. Robinson and K. A. Holbrook, *Unimolecular Reaction Rates*, John Wiley and Sons, New York, 1972.

60. D. W. Oxtoby and S. A. Rice, *J. Chem. Phys.* **65** (1976) 1676.

61. D. L. Bunker and W. L. Hase, *J. Chem. Phys.* **59** (1973) 4621.

62. A. Kaldor and M. Berry, *Symposium on Laser Chemistry*, M.I.T. Chemical Sciences/Industry Forum, June 1977.

63. J. I. Steinfeld and M. S. Wrighton, "The Laser Revolution in Energy-Related Chemistry," Energy-Related General Research Program Office, National Science Foundation, Washington, D.C., May 1976.

64. J. T. de Maleissye, F. Lempereur, and C. Marsal, *Chem. Phys. Lett.* **42** (1976) 472.

65. R. Zitter and D. F. Koster, *J. Am. Chem. Soc.* **100** (1978) 2265.

66. V. S. Letokhov and C. B. Moore, *Sov. J. Quant. Electron.* **6** (1976) 259.

67. R. V. Ambartzumian, N. V. Chekalin, V. S. Doljikov, V. S. Letokhov, and N. V. Lokhuan, *J. Photochem.* **6** (1976/77) 55.

68. A. Yogev and R. M. J. Benmair, *Chem. Phys. Lett.* **46** (1977) 296.

69. I. Glatt and A. Yogev, *J. Am. Chem. Soc.* **98** (1976) 7087.

70. J. J. Ritter, *J. Am. Chem. Soc.* **100** (1978) 2441.

71. H. R. Bachmann, H. Nöth, R. Rinck, and K. L. Kompa, *Chem. Phys. Lett.* **29** (1974) 624.

72. J. W. Robinson, P. J. Moses, and P. M. Boyd, *Spectrosc. Lett.* **7** (1974) 395.

73. W. E. Farneth, G. Flynn, R. Slater, and N. J. Turro, *J. Am. Chem. Soc.* **98** (1976) 7877.

74. D. Garcia and P. M. Keehn, private communication, August 1977.

75. J. C. Bellows and F. K. Fong, *J. Chem. Phys.* **63** (1975) 3035.

76. A. Yogev and R. M. J. Lowenstein-Benmair, *J. Am. Chem. Soc.* **95** (1973) 8487.

77. S. Kimel and S. Speiser, *Chem. Rev.* **77** (1977) 437.

78. J. N. Butler, *J. Am. Chem. Soc.* **84** (1962) 1343.

79. J.-B. Wiel and F. Rouessac, *J. Chem. Soc., Chem. Comm.* (1976) 446.

80. A. Yogev, R. M. J. Lowenstein, and D. Amar, *J. Am. Chem. Soc.* **94** (1972) 1091.

81. A. K. Petrov, A. N. Micheev, V. N. Sidelnikov, and V. N. Molin, *Dokl. Akad. Nauk SSSR* **212** (1973) 915.

82. H. R. Bachmann, R. Rinck, H. Nöth, and K. L. Kompa, *Chem. Phys. Lett.* **45** (1977) 169.

83. N. V. Karlov, Yu. N. Petrov, A. M. Prokhorov, and O. M. Stel'makh, *JETP Lett.* **11** (1970) 135.

84. N. J. Turro, P. Lechtken, N. E. Shore, G. Shuster, H.-C. Steinmeyer, and A. Yekta, *Acc. Chem. Res.* **7** (1974) 97.

85. H. R. Bachmann, F. Bachmann, K. L. Kompa, H. Nöth, and R. Rinck, *Chem. Ber.* **109** (1976) 3331.

86. E. R. Long, S. H. Bauer, and T. Manuccia, *J. Phys. Chem.* **79** (1975) 545.

87. H. R. Bachmann, H. Nöth, R. Rinck, and K. L. Kompa, *Chem. Phys. Lett.* **33** (1975) 261.

88. J. L. Lyman and R. J. Jensen, *Chem. Phys. Lett.* **13** (1972) 421.

89. I. N. Kal'vina, V. F. Moskalenko, E. P. Ostapchenko, L. L. Pavlovskii, T. V. Protsenko, and V. I. Rychkov, *Sov. J. Quant. Electron.* **4** (1975) 1285.

90. S. G. Il'yasov, I. N. Kal'vina, G. A. Kyulyan, V. F. Moskalenko, and E. P. Ostapchenko, *Sov. J. Quant. Electron.* **4** (1975) 1287.

91. E. K. Karlova, N. V. Karlov, G. P. Kuz'min, B. N. Laskorin, A. M. Prokhorov, N. P.

Stupin, and L. B. Shurmel', *JETP Lett.* **22** (1975) 222.
92. J. J. Ritter and S. M. Freund, *J. Chem. Soc., Chem. Comm.* (1976) 811.
93. J. J. Ritter, *J. Am. Chem. Soc.* **100** (1978) 2441.
94. S. Pinchas and I. Laulicht, *Infrared Spectra of Labeled Compounds*, Academic Press, New York, 1971.
95. R. V. Ambartsumian, N. P. Furzikov, Yu. A. Gorokhov, V. S. Letokhov, G. N. Makarov, and A. A. Puretzky, *Opt. Lett.* **1** (1977) 22.
96. American Institute of Physics, *Handbook*, 3rd ed., McGraw-Hill, Inc., New York, 1972.
97. Chemical Rubber Company, *Handbook of Lasers*, CRC Press, Cleveland, Ohio, 1971.
98. C. K. N. Patel, *Appl. Phys. Lett.* **7** (1965) 246.
99. Lumonics Research, Ltd., technical data (1976) for model TEA 203 laser, P.O. Box 1800, Kanata, Ont., Canada.
100. T. F. Deutsch, *IEEE J. Quant. Electron.* **3** (1967) 419.
101. P. Carruthers and M. M. Nieto, *Am. J. Phys.* **33** (1965) 537.
102. F. M. Lussier, *Laser Focus*, **12** (December 1976) 47.
103. *Laser Focus Buyers' Guide* 1977, p. 214.
104. D. E. McCarthy, *Appl. Opt.*, **2** (1963) 591 and 596.
105. D. E. McCarthy, *Appl. Opt.*, **4** (1965) 317 and 507.
106. Harshaw Optical Crystals, Harshaw Chemical Co (1967).
107. J. E. Stewart, *Infrared Spectroscopy: Experimental Methods and Techniques*, Marcel Dekker, Inc., New York, 1970.
108. L. S. Kassel, "The Kinetics of Homogeneous Gas Reactions," A. C. S. Monograph No. 57, New York, 1932, Sect. 5.3.
109. J. Bastiaens, D. De Keuster, and X. de Hemptinne, *Bull. Soc. Chim. Belg.* **85** (1976) 833.

Index

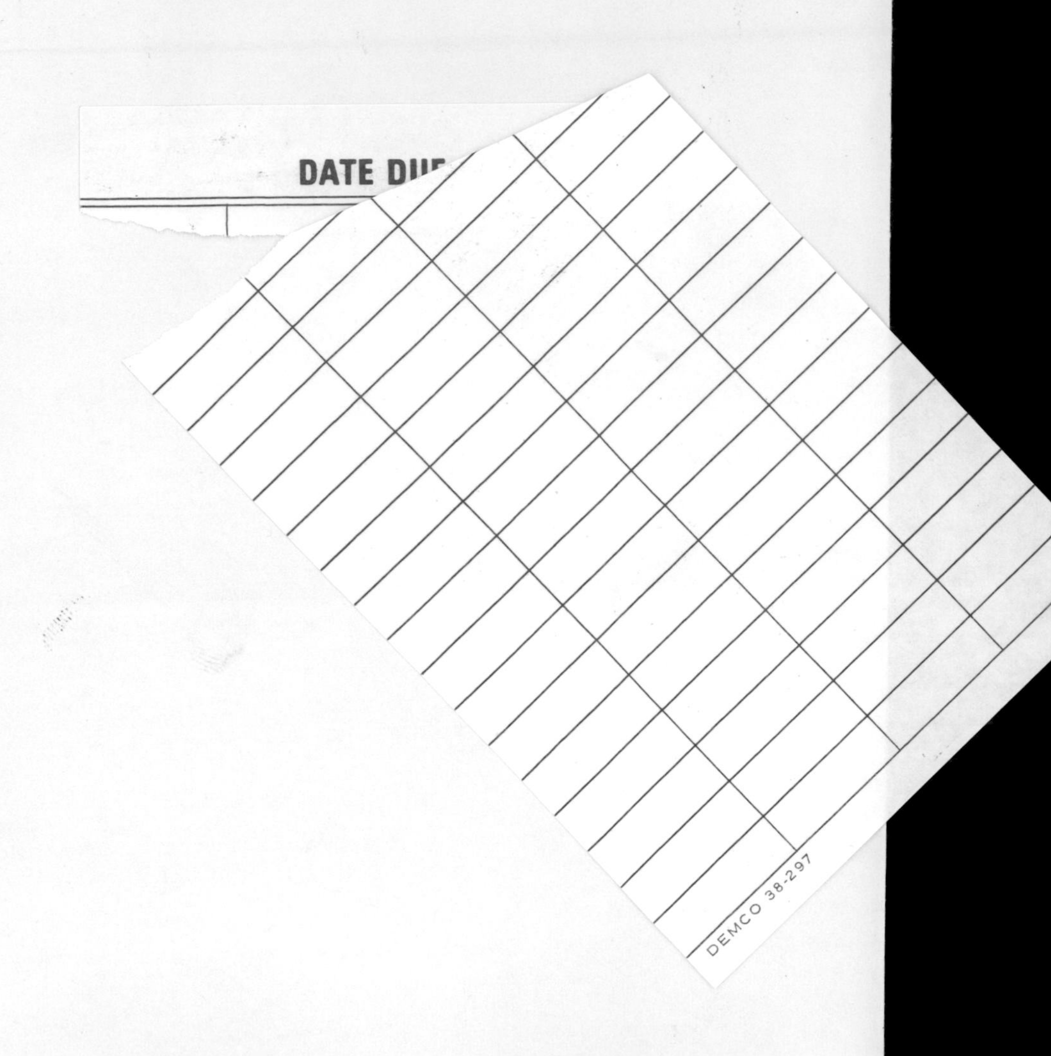